PIONEERS OF PLANT STUDY

PIONEERS OF PLANT
STUDY

CAROLUS LINNAEUS (1707-1778).

Frontispiece.

Pioneers of Plant Study.

PIONEERS OF PLANT STUDY

BY

ELLISON HAWKS

This book was originally planned, and some parts
of it written, in collaboration with the late

GEORGE SIMONDS BOULGER

Essay Index Reprint Series

BOOKS FOR LIBRARIES PRESS
FREEPORT, NEW YORK

First Published 1928
Reprinted 1969

STANDARD BOOK NUMBER:
8369-1139-3

LIBRARY OF CONGRESS CATALOG CARD NUMBER:
75-86759

PRINTED IN THE UNITED STATES OF AMERICA

PREFACE

THIS book makes no attempt to be a complete history of plant study. Had this been its object, we should have had to deal with a vast amount of detail that can be of little interest except to the advanced student. The treatment of the subject is necessarily episodic, and there are not a few interesting episodes that we have been compelled, by exigencies of space, to pass by. We should like to have spoken of the Tradescants and Ashmole and of the first museum in England; of the Sherards and Bobarts and the history of the Oxford Garden; of Goethe and his poetical suggestion of serial metamorphosis; of the plant-breeding Abbot Mendel and his far-reaching experiments; of Darwin's contributions to our knowledge of insect-pollination, and to the physiology of climbing and of insectivorous plants; and of the manner in which our knowledge of fossil plants has grown of late, and how it has thrown light on the structure of living forms.

Great as have been the victories of the human intelligence over the forces of the vegetable world, the problems solved seem only to point the way to further questions that yet await solution. As in all other sciences, the more we learn, the wider does the path open out, so that we see ever more before us. How far we yet are from a complete knowledge even of the Flowering Plants of the world may be gauged from the fact that within thirty years of the issue of the earlier volumes of the *Flora of Tropical Africa*, about the year 1870, an increase of more than 80 per cent. in the number of species had been described in genera that had been revised. While this is probably typical of all Tropical African groups, our knowledge of Tropical South American species is probably at least equally imperfect.

The Cryptogams of the world are even less well known than the Flowering Plants.

So many regions are even yet so incompletely explored that we are ignorant of the mere facts of geographical distribution, and far less have we been able to trace the routes by which plants have migrated, or their places of origin. Plant classification having been based upon external appearance, we are only now beginning to learn how differences of microscopic structure correspond with external differences, very few plants having been at all thoroughly examined with the microscope.

Although the broad facts of latitudinal and altitudinal distribution, and the requirements of the species we cultivate as to temperature and moisture, have been noticed, we are yet very far from knowing all the complex relations of each species to the external conditions that determine its living, flourishing, maturing, and succeeding in competition in any particular spot. We now know the soil to be not merely an inert mixture of more or less soluble mineral matters. It is teeming with varied and competing micro-organisms, each of which plays a most important, but most imperfectly investigated, rôle in rendering various substances available as food material for the higher plants, or making the soil more suitable for them. The effects of soil temperatures and air temperatures at the different seasons of the year upon the germination, growth, flowering, and fruiting of different species; those of fluctuations in the water supply, or in the intensity of light, or of the presence or absence of various ingredients in the soil, are only some of the more important physiological problems that are as yet very far from complete solution.

In the space at our disposal it has only been possible briefly to indicate how far men have been able to catalogue the plants of the world; to trace their distribution, to investigate their structure, conditions of life, physiology, and cultivation, and to discover their many uses in the arts of life—uses that even now are still very small and

probably apply to less than two per cent. of known species. It
has been our endeavour to trace the work of the pioneers
who opened up to the world some of the treasures of what is
vaguely termed the vegetable kingdom, and this we have
done from the earliest times through the ages to the 19th
century. It is hoped that in the near future it will be
possible to publish a companion volume, dealing with the
" Pioneers of Plant Culture " and advancing the story of
plant knowledge from the early 19th century to the present
day, thus completing the story commenced in this first
volume.

It only remains to add that, as indicated on the title page,
the present volume—although the conception of the undersigned—was planned, and some parts of it written, in
collaboration with the late G. S. Boulger, whose name is
well known to all botanists. The work was interrupted
first—as far as the undersigned was concerned—by active
service during the War, and later by illness in Professor
Boulger's household. When, six years ago, the Professor
died very suddenly, it was decided to proceed with the book
on the lines originally planned. Owing to the considerable
amount of research necessary, however, its completion has
taken longer than was anticipated. It is hoped that in some
measure the book will fill a vacant space in botanical
literature, for it covers a wide and—as yet—a little trodden
field.

In conclusion I must express my indebtedness to Mr.
W. H. McCormick and to Mr. J. A. Notman for kindly
reading through the proofs.

ELLISON HAWKS.

" Arcetri,"
   Dudlow Lane,
      Liverpool.
         April, 1928.

CONTENTS

CONTENTS

LIST OF ILLUSTRATIONS

PIONEERS OF PLANT STUDY

CHAPTER I

IN EXPLANATION OF TECHNICAL TERMS

ALTHOUGH the main purpose of this volume is historical and biographical, it may be as well to prefix a brief explanatory introduction in order that those who are not botanists may follow the technical terms that inevitably must occur.

Botany, as the science of plants in all their aspects, may be broadly divided into three sections: (1) Pure, (2) Mixed, and (3) Applied.

(1) The pure science deals with the structure, or anatomy, and the functions, or physiology, of plants. Anatomical study, when comparative with reference to fundamental types or series of variously modified structures, is known as Morphology. Typical structures are termed members, and their various modifications are said to be homologous to one another.

(2) Mixed Botany deals with the distribution of plants in time and space; Palæobotany, the study of fossil plants being mixed with geology, and Phytogeography with geography.

(3) Applied or Economic Botany deals with the uses of plants. These may be as food of man (or of his domesticated animals), as medicine, or in the form of timber, or for clothing, or other manufactures.

Almost all plants are built up of structural units known as cells, or of structures derived from the modification or fusion of cells, such as wood-fibres or vessels. The study of this minute or microscopic anatomy is known as Histology, the science of tissues, a tissue being a combination of similar cells or vessels.

As everyone knows, plants differ enormously in size, structure, and physiology. They comprise a prodigious number of kinds or species, with more or less obvious likenesses between them, suggesting their common origin from a certain number of types. A generally recognised system of nomenclature is obviously necessary for their ready identification, and their mere multitude compels us to adopt some grouping or classification, the study of which is known as Taxonomy.

It is now agreed that every species shall have a Latin name that is binary or binominal. This consists of two words, the first of which is the generic name, which is shared by the other related species of the same genus or wider group, the second, or specific, name being peculiar to the particular species. The latter is followed by the name (usually abbreviated) of the authority—the botanist who first applied that binary name to the plant. Thus in *Rosa canina* L., although the generic name (*Rosa*) is shared by other species of Rose, *canina* belongs to the Dog Rose only, while L. indicates that the name was given by Linnæus, the first authority to regulate this system of binary nomenclature.

A mere grouping of plants for convenience according to some easily-recognised character, or set of characters, is termed an artificial classification. An example of this is the alphabetical arrangement of names in an index, or the ancient division by size into trees, shrubs, and herbs. It is now generally admitted, however, that it is possible to arrange plants, by the consideration of all their structural characters, in a natural system representing their relationship to one another. Such an arrangement is, in fact, equivalent to a pedigree, each successively wider group representing a more remote common ancestry.

Botanists fully realise that the Natural System is the sum of their knowledge and can never be considered as absolutely completed. Species are placed together in Genera, Genera in Families, Families in Orders, Orders in Classes, and Classes in Sub-kingdoms, some intermediate groupings being also employed. The Latin names of the Families end in *-æ* or *-aceæ*, being adjectives agreeing with the word *plantæ*. Thus *Ranunculaceæ* means those plants that are ranunculaceous, *i.e.*, which resemble the genus *Ranunculus*. Sometimes the Family name merely implies the possession of some character in common. *Compositæ*, for instance, are plants with a particular form of crowded inflorescence or grouping of small flowers or florets, known as a *capitulum*. *Umbelliferæ* are plants in which the flowers are borne in an *umbel*, *i.e.*, on stalks all radiating from one point. *Labiatæ* are plants with a *bilabiate*, or two-lipped corolla.

Among groups of plants generally recognised as " natural " are Fungi, Algæ, Mosses, Ferns, and Flowering Plants. The first four groups have long been known as *Cryptogamia*, because of the minute or hidden character of the dust-like spores and other organs by which they reproduce themselves. Flowering or Seed-bearing Plants are known as *Phanerogamia* or *Spermatophyta*. At the present time upwards of 100,000 species of Flowering Plants have been described by botanists, and nearly as many cryptogams.

The more complex or highly-organised plants that are now characterised by bearing the elaborate, though compact, detachable reproductive structure that we know as a seed, fall into two well-marked but unequal groups. In the first, which are all woody plants (and of which our needle-leaved cone-bearing Pines and Firs are the most familiar examples), the seeds are naked, and they are accordingly known as Gymnosperms. Their seeds are generally borne on the surface of scales that overlap, in the structure known as a cone. The second, and by far the more numerous and varied division, is known as Angiospermia, because the seed is enclosed by one or more organs of the leaf type. These organs, known as carpels, form a

hollow structure, the ovary, and this afterwards constitutes
the main portion of the fruit.

Angiosperms are divided into two Classes, named
Dicotyledons and Monocotyledons respectively. They have
two seed-leaves or cotyledons to the embryo in the seed, or
only a single cotyledon, and are also generally contrasted
with one another in the structure of their roots, stems, leaves,
and flowers.

The organs of a plant may be described as being either
vegetative, when they serve in the nourishment and development of the individual, or reproductive, when they serve in
the multiplication of the species.

Among Spermatophytes the chief vegetative organs are
the root, stem, and leaves, the chief reproductive parts
being the flower, fruit, and seed.

Gymnosperms and Dicotyledons have commonly an
elongated tapering or tap-root, branched or unbranched.

Monocotyledons have a bunch of unbranched or fibrous roots.

The main function of all roots is the taking in of liquid
food-material, in which gaseous and mineral matters are
dissolved.

Gymnosperms and Dicotyledons agree also in the general
structure of their stems when they become woody, while
those of Monocotyledons are markedly different. Trees
belonging to the former groups have a separable bark and
increase in diameter by the formation of new cells in a
layer immediately beneath the bark. This layer is termed
the *cambium*, and as it is external to the wood of the stem
this type of stem is called exogenous.

When young the stem is entirely made up of cellular
tissue (*protomeristem*). The cells are at first not longer than
they are broad, such tissue being known as *parenchyma*, but
the cells soon become elongated, mainly in the direction
in which the stem grows. Towards the centre of the stem
some cellular tissue undergoes little change and is long
recognisable as the soft pith. Between this and the bark
in the exogenous stem there originates a circle of structures,

each of which appears wedge-shaped in a cross-section of the stem. These are known as the fibro-vascular bundles, and in them the walls of the cells become thickened and woody. Some of the vertical rows of cells may become fused by the absorption of their partition walls into long tubes, known as vessels or *tracheæ*.

The thickening of the first-formed vessels is generally a spiral band that allows of a considerable elongation of the vessel. Between these wedge-shaped bundles the cellular structures of the stem elongate mainly in a radial direction, forming what are known as the rays or silver-grain. The bundles are added to, and the whole stem increases in diameter, by the growth in the zone of cambium, each season's growth forming one of the annual rings of wood so well known as distinctive of the exogenous stem.

The wood of Gymnosperms is simpler than that of Dicotyledons and consists mainly of long wood fibres, known as *tracheids*. They have no vessels but occasional relatively large ducts containing turpentine and known as resin-passages.

Dicotyledonous wood may consist of fibres, tracheids, cells and vessels in varying proportions. The vessels may be wide and open for the upward passage of water to the leaves, or they may be blocked by cellular ingrowths, known as tyloses. The older rings of wood may become choked, physiologically dead and often discoloured, when they are known as the heart-wood, the younger outer layers being called sapwood or alburnum.

The stems of Monocotyledons, on the other hand, such as Palms and Bamboos, are very different. Their bundles are more numerous, are not in a ring and have no cambium. There is accordingly neither distinct pith, separable bark, nor annual rings. The stem is hardest towards the outside, is seldom branched, and maintains a very uniform diameter throughout.

There are many special modifications of the stem, among the most important of which are those adapted to the underground life of herbaceous plants that are perennial but die down annually.   Such are the more or less elongated and more or less fleshy rhizome, often unfortunately called a root-stock (exemplified in the Lily-of-the-Valley, Solomon's-seal, Iris and Couch-grass), and the much-shortened form known as a bulb, giving off roots below and entirely enveloped above by overlapping leaf-scales.

Stems are in general distinguished from roots by bearing leaf-buds, the bud being a rudimentary stem or branch, the growing-point of which is enclosed by leaf-scales.   The leaves spring in definite order from the stem, which has often an internal structure in the region whence the leaves originate different from that where they do not.   In Bamboos, Wheat, and other Grasses, for instance, the stem is solid at the base of each leaf, but hollow between the leaves. These solid regions are termed nodes, and the regions between them are known as the internodes.   The slender stem of Grasses, with these long hollow internodes, is termed a culm.

Physiologically the leaf is the most important of the vegetative organs.   It originates from the outer tissues of the stem and is, therefore, exogenous.   Through it run what are variously known as veins, ribs, or nerves, the names having been given them in a futile attempt to explain their functions by analogy with the structures of animals. These are made up of vessels, continuous with those of the stem, by which water (with nutritive substances in solution) travels upward from the root, and food travels back to the stem.   The leaf is always covered by a cell-layer known as the epidermis, which in stems only remains usually for the first year.   This is pierced, especially on the underside of the leaf, with numerous openings known as *stomata* (Lat. " little mouths "), by which the excess of water that has served to carry up food-material from the root is transpired.   Through the *stomata*, the opening of which is under physiological control, air also passes into the plant.   This air is the source of the small amount of oxygen required for breathing or respiration and also of carbon dioxide, the main source of the carbonaceous food-material of the plant.

The cellular tissue in the interior of the leaf (mesophyll) contains (except in the case of brown parasites and saprophytes [1]) the characteristic green substance known as chlorophyll. The special function of this substance, known as the chlorophyllian action, or photosynthesis, is to decompose the carbon dioxide, under the influence of light, and to build it up into sugar and starch, which is one of the chief stages in the process of assimilation or food-manufacture. So that they may best perform these three functions of transpiration, respiration, and photosynthesis it is desirable that the leaves should be exposed to air and light. For this reason they are naturally variously arranged on the stem or branches, so as to overlap as little as possible. If they are given off singly from successive nodes they are termed scattered; if two or more together, whorled. Scattered leaves exactly on opposite sides of the stem, forming two vertical rows, are termed alternate. Leaves in whorls of two (*i.e.* in pairs) are termed opposite.

The leaf may consist only of an expanded blade, when it is termed sessile. More commonly, however, it has also a narrowed stalk or petiole, when it is called petiolate. When a leaf has only one continuous blade it is termed simple, even if deeply but not completely divided. When it is made up of several distinct blades or leaflets it is termed compound. The leaflets of compound leaves may all radiate from the apex of a common petiole (as in the Horse-chestnut), when they are said to be digitate or palmate. When they are given off on either side of an elongated stalk, thus resembling barbs of a feather, they are called pinnate. In this latter case there is usually an odd or terminal leaflet (as in the Rose), when the leaf is termed imparipinnate, or unequally pinnate. If there is no terminal leaflet it is said to be paripinnate.

[1] Plants living on decaying organic matter.

B

There are no characters that serve so readily to discriminate between closely related species as those of the leaf. Descriptive botany, therefore, has a very large number of more or less precisely defined terms referring to all the external features of this organ, its texture, margin, base, apex, outline, veining (venation), etc. Plants adapted to drought, for instance, known technically as *xerophytic*, may have the fleshy leaves with a thick epidermis and few *stomata*, such as we find in the Houseleek and in a large number of South African plants. Monocotyledons have usually an unbroken or entire margin to their leaves, while many Dicotyledons have them toothed (serrate) or otherwise notched or divided. The prominent veins in the Monocotyledonous leaf run parallel with the margin, while cross veins are scarcely visible. In Dicotyledons, however, the veins repeatedly branch, forming a complex network.

In Temperate latitudes, with sharply contrasted seasons, trees and shrubs usually shed their foliage in autumn, and are known accordingly as deciduous. In the forests of the Equatorial zone of constant precipitation, where, on the other hand, there is little or no seasonal change, the trees are evergreen. That is, they have always some leaves, but the individual leaves need not necessarily have any longer duration than those of deciduous trees.

The branch, or system of branches, bearing the flowers is known as the inflorescence. If it rises directly from an underground stem, free, or nearly free, from foliage-leaves (as in Tulip, Primrose, or Lily-of-the-Valley) it is termed a scape.

The main distinctive characters among inflorescences are whether the lower or outer flowers or those in the centre are the first to expand. That is, whether the succession is centripetal or centrifugal, and whether the flowers are borne on simple or branching stalks or are crowded together without any stalks. An elongated axis with sessile flowers is termed a spike (as in *Plantago major* L., the Great

Plantain), but when a spike falls off whole (as does the male inflorescence of the Hazel) it is known as a catkin.

An elongated axis bearing centripetal stalked flowers (as in the Lily-of-the-Valley) is termed a raceme; but if the stalks of the lower flowers are so elongated as to bring the blossoms up to one level, thus rendering them more conspicuous in the mass (as in the Wallflower) it is called a corymb. We have already seen that when the main axis, or peduncle, ends in branches all radiating from one point (as in the Ivy) it is known as an umbel. When these branches are absent, so that all the flowers are massed together sessile on the expanded apex of the peduncle, we have the *capitulum* or head.

This type of inflorescence characterises the largest Family of Flowering Plants, the *Compositæ*, which comprises more than 10,000 species—a tenth of all known Spermatophytes. The small crowded flowers are known as florets. They may be all alike and tubular (as in the Groundsel), all alike and strap-shape (as in the Dandelion), or those in the centre may be tubular and the outer ones strap-shaped (as in the Daisy). In this last case the centre florets are termed the disk and the outer ones the ray-florets. Disk and ray may be both of the same colour (as in most Cinerarias) or of different colours (as in the Daisy).

From the point of view of its development the flower may be described as an unbranched and generally abbreviated axis bearing a series of foliar or leaf-like organs. These are specially modified in form, structure, texture, and colour to serve more or less directly in the production of seed. On the floral axis, known as the receptacle or *thalamus*, the floral leaves may be arranged either spirally, in successive circles or whorls, or partly in one way, partly in the other. If in each whorl the leaves are alike, the flower is termed regular; if they are unequal, it is irregular. (The Buttercup and the Lily are examples of the former and the Orchids and Salvias of the latter.) The number of floral leaves in each whorl is commonly three among Monocotyledons, and five, four, or two among Dicotyledons. As a rule the leaves in the successive whorls alternate—that is to say, the leaves in any one whorl are over the spaces between the leaves of the whorl below.

In a typical fully-developed flower there will be four kinds of leaves—sepals, petals, stamens, and carpels. The sepals, forming collectively the calyx, are commonly green and leaf-like, serving mainly as a protection to the other floral organs in the bud stage. They may be carried up on a tubular base, when they are said to be gamosepalous (as in the Primrose). The petals, forming collectively the corolla, are commonly delicate in texture, not green but white or coloured and often fragrant, their common function being the attraction of insects to the flower. Like the sepals, the petals may be more or less united below, or gamopetalous, a condition that also occurs in the Primrose. The tube of the corolla often serves to hold the nectar or honey that acts as the bait to insect visitors. Corollas vary much in form, among the more striking of regular types being the star-like or stellate form in Pimpernels and the bell-shaped in *Campanula*.

Calyx and corolla are known together as the floral envelope or perianth, the latter term being more commonly employed when one whorl is absent or when the leaves of both are alike. The exceptional tubular outgrowth from the perianth in Narcissus, and the similarly situated circle of coloured hair-like structures in the Passion-flowers, are known as the *corona* or coronet. If either calyx or corolla be absent, the flower is termed incomplete, and asepalous or apetalous.

As being more indispensable for the formation of seed than either sepals or petals, the stamens and carpels are termed the essential organs. In many cases, however, they do not occur in the same flower, or even on the same plant. A flower containing both stamens and carpels is known as perfect.

When they occur in distinct flowers, the flowers are imperfect or diclinous and either staminate or carpellate, as the case may be. If staminate and carpellate flowers occur on the same tree (as in Hazel, Oak, or Alder) the plant is termed monœcious. If on distinct plants (as in Willows, Poplars, or White Bryony), the species is said to be diœcious.

When the flower is structurally perfect, it by no means follows that its ovules, or immature seeds, will be fertilised by the pollen of its own stamens, or be self-pollinated, as it is termed. Very often the stamens and carpels will be found to mature at slightly different times, when they are known as dichogamous. On the other hand, the stamens may throw their pollen away from the carpels. Such arrangements favour fertilisation by pollen brought by wind or insects from another flower—cross-pollination, as it is termed. This process is apparently favourable to the race, introducing seedling variations that may prove better fitted for their surroundings.

The stamens vary in number, being commonly twice the number of the petals and in two alternating whorls, or equal in number to the petals, or indefinitely numerous, or even fewer than the petals. Their number, relative lengths, and union formed the primary basis of the Linnæan artificial system of classification. Each stamen consists typically of a slender stalk or filament, and a rounded or elongated, generally two-lobed structure. This latter is known as the anther, and when ripe bursts open to discharge its fine dust-like spores or pollen-grains. (To prevent the self-pollination and consequent rapid withering of Lilies it is a common practice with florists to remove the anthers from the newly-opened blossom.)

The carpels vary in number, being indefinitely numerous. They are spirally arranged where the floral axis is elongated (as in Magnolia, the Mouse-tail (*Myosurus*), *Ranunculus*, or *Rubus*), but are more commonly few and arranged in one whorl of five, four, three or two, or even solitary (as in the

Pea Family, *Leguminosæ*). The carpel, or carpellary leaf, has little or no stalk but widens out at its base, either to form or to contribute to the formation of the hollow cavity known as the ovary. Upwards the carpel may taper into a narrower rod-like portion or style, terminating above in a more or less expanded surface, the stigma. This, when mature, is sticky with a sugary exudation in which the pollen-grains are retained and nourished. The carpels may be free from one another throughout, or apocarpous (as in *Sedum*), or they may be variously united or syncarpous, either in the ovarian region or also in that of the styles. The number of styles was used by Linnæus to characterise the Orders into which he subdivided the Classes founded on the stamens. Internally the ovary may be either one-chambered or divided up into chambers usually equalling the carpels in number.

A very important floral character, first insisted upon by the de Jussieus, is what is termed the insertion of the petals, stamens, and ovary with reference to the receptacle. Where the receptacle tapers, and sepals, petals, and stamens spring successively from it obviously below the carpels (as in a Buttercup) the calyx is termed inferior, the corolla and stamens *hypogynous* (*i.e.* below the ovary), and the carpels superior. Where the floral axis is arrested in its elongation it may spread out laterally below its apex either into a disk (as in *Rubus*) or into a cup (as in *Prunus* and *Rosa*). The calyx then is still inferior and the carpels superior, but the petals and stamens are termed *perigynous*, being borne in a ring round the carpels. If this receptacular tube appears to adhere to the sides of the carpels so that calyx, corolla, and stamens appear to spring from the top of the ovary and the ovary is visible as a swelling below the other parts of the flower (as in Narcissus, Fuchsia, or Ivy) the calyx is termed superior, the ovary inferior, and corolla and stamens *epigynous*.

Within the ovary, the ovules or immature seeds vary greatly in number from one, two, or four to an indefinite number. They spring usually from the enlarged inrolled margins of the carpellary leaves that form a spongy mass known as the *placenta*. Where two carpels meet there is thus often a double row of ovules. These rows of ovules may run down the inside of the outer wall of a one-chambered ovary (as in the Violet) which is termed parietal placentation. Where the margins of the carpellary leaves are rolled in, so as to meet at the centre and form a many-chambered ovary (as in Flax) the placentation is central. Less commonly, as in Poppies, the ovules are scattered over the inner surface of the carpels, and this is known as superficial placentation.

Each ovule originates as a minute cellular out-growth on the carpel, to be known later as the nucellus. A small vein or vascular bundle generally extends from the carpel to the base of each ovule, its purpose being to convey nourishment to it. From the base of this rudimentary ovule, or nucellus, one or two coats, known as secundine and primine, grow up. They enclose the nucellus with the exception of one spot, where a minute aperture, the micropyle, serves to permit the entrance of the pollen-tube. Within the nucellus a cavity, known as the embryo-sac, is formed, and in this the ovum or egg-cell originates by successive subdivisions of the central body, known as the nucleus of the embryo-sac.

When the ripe anther bursts, the escaping pollen-grains either fall on to the stigma of the same flower (self-pollination), or more commonly are carried by wind or insect visitors to that of another flower, possibly on another plant of the same species. Nourished by the secretion of the stigma, the pollen-grains burst their outer coats at certain points and protrude tube-like outgrowths, known as pollen-tubes. These grow down the central canal of the style—which may be empty, but is usually filled with conducting tissue—into the ovary. There, guided by the soft tissue of the placenta, they find their way into the micropyles of the ovules, and the first to enter any micropyle also penetrates the outer tissue of the nucellus and the embryo-sac. The fusion of a nucleus from the pollen-tube with that of the ovum or egg-cell constitutes fertilisation, or the setting of the seed.

Great changes then begin. The stamens and corolla generally wither, shrivel up, and disappear. The ovary enlarges to form a fruit, and its contained ovules, now known as seeds, not only enlarge but also undergo various structural changes.

As the enlarged fruit ripens it may become either dry or fleshy. In the former case, if it contains more than one seed, it commonly splits open or dehisces in certain definite directions, sometimes (as in Gorse) elastically so as to scatter the seeds. Where there is an inferior ovary, and in some other cases, the receptacle, the calyx, or other structures may contribute to the building up of one of the varied types of fruit. Where the fruit is formed by the ovary alone, however, its walls—apart from the seeds that they enclose—are known as the pericarp. When (as in Cherry, Plum, or Peach) this becomes fleshy, it is often clearly divisible into three layers—an outer skin, or epicarp (which is polished in the Cherry, covered with a waxy bloom in the Plum, and downy in the Peach), a middle fleshy layer, or mesocarp, and an inmost layer, often stony, known as the endocarp.

Meantime, while the acid juices of the fruit may give place to sugar, and the green chlorophyll of its outer layers redden in the sun, changes both external and internal occur in the young seeds. They form one or two coats, the exterior one, or testa, being commonly thick, opaque, leathery, and bitter; the inner, or endopleura, thin and ivory-white. In fruits that dehisce, the testa is often brightly coloured (as in Beans), in those that do not (such as Almonds, Nuts, or Apples) it is commonly brown. In some dehiscent fruits (such as the Nutmeg) an extra fleshy-coloured covering, or aril, grows up outside the testa. This acts as an attraction to birds, which then swallow the seeds whole and disperse them without digesting them. The scarlet aril of the Nutmeg is known as Mace.

The fertilised ovum gives rise to an embryo or young plant. The rest of the embryo-sac becomes filled with cellular tissue, storing food for the young plant—a tissue known as endosperm. A similar food-store may also originate in the nucellus outside the embryo-sac, and is known as the perisperm. When both, or either, of these tissues remain unabsorbed when the seed is ripe it is termed albuminous; but in other cases all the nutriment will have been transferred into the embryo itself. In these cases the ripe seed will consist solely of coats and embryo. In a Dicotyledon the embryo will consist of two more or less hemispherical seed-leaves or cotyledons, a rudimentary tap-root or radicle, and an even more rudimentary bud or shoot, the plumule. In the seed of the Almond, for instance, within the brown bitter testa is the ivory-white endopleura. The contained embryo falls into two planoconvex halves or " split almonds," sunk in the plane inner surfaces of which may be detected the radicle and plumule. Such a seed, with its sole food-store in its cotyledons, is termed exalbuminous.

The study of the relation between such periodic phenomena of plant life as the sprouting or germination of seeds, the unfolding of leaves and flowers, or the ripening of fruit, and the seasons of the year—a study mixed with the science of meteorology—has been named Phenology.

The structural characters of plants are largely hereditary, being transmitted by seed or true to seed. Many features, and especially in the vegetative organs, are liable to considerable modification, however, by the action of the surrounding circumstances of their growth. These circumstances collectively are termed the habitat of the plant. The study of the plants associated together in certain habitats, and of the reaction of their surroundings upon them, is known as Ecology. Plants not otherwise related which happen to grow in water on the sea-shore, or in some other situation where but little fresh water is procurable by their roots, will be modified in very similar ways in leaf and stem, and are known as hydrophytes, halophytes, and xerophytes respectively.

CHAPTER II

THE PLANTS OF ANCIENT EGYPT

IT is by no means easy to identify with any certainty the comparatively few species of plants represented in the ancient monuments of Egypt. Even less is it possible to determine those alluded to in the hieroglyphic inscriptions. But in the dry soil and atmosphere of Egypt many actual specimens of plants have been preserved to us. Some, no doubt, were set apart for food for the workmen engaged in building the ancient pyramids or temples. Others were twisted into garlands offered at the tombs of the dead, or remaining from solemn funeral feasts. We know nothing of the individual men who collected or cultivated these plants, but from the remains it is possible to form some idea of the fields, kitchen-gardens, and orchards of Egypt, and of the plants grown in them before the time of Abraham (2400–2200 B.C.).

At Kahun, Professor Flinders Petrie discovered (in 1889), among remains belonging to the Twelfth Dynasty, a quantity of grains of Barley smaller than those now grown in Egypt. Among them were the seeds of various weeds, such as the Egyptian Clover (*Trifolium alexandrinum* Linné), the spiny Medick (*Medicago denticulata* Linné), the scarlet Poppy (*Papaver Rhœas* Linné), an Oat (*Avena strigosa* Schreber), a garden Pea (*Pisum arvense* Linné), and the Flax then cultivated (*Linum humile* Miller). Thus it seems that the Egyptian fields of that date had much the same weeds as those of to-day.

At a later, but still ancient, period, Mr. Percy Newberry has found the Egyptian Clover, originally a native of Asia Minor, to have been attacked, as it is to-day, by the Arabian

17

Dodder (*Cuscuta arabica* Linné). Beans, Radishes, and the leaves and stems of the Cucumber also were found at Kahun, recalling the statement of Herodotus that Radishes, Lentils, Onions, and Garlic were supplied to the builders of the Great Pyramid. The fruits and seeds of the Carob (*Ceratonia Siliqua* Linné), the Dom and Dellach Palms (*Hyphœne thebaica* Martius and *H. Argun* Martius), the Jujube (*Zizyphus Spina-Christi* Willdenow), and the Sycamore Fig (*Ficus Sycomorus* Linné) were among those found in this most ancient collection. Boxes made of the wood of the Fig tree and many articles of the wood of *Acacia nilotica* Delille, tell of the long persistence of such uses for common plants.

A few fragments of plants have been found accidentally preserved in the sun-baked bricks of ancient times, in addition, that is, to the straw of Barley or Wheat that was used to bind them. By far the most remarkable collection, however, was that found in 1888 and 1889 by Professor Petrie in a Greco-Roman cemetery at Hawara in the Fayum. Here, among a quantity of fruits, seeds and leaves, probably the remains of funeral banquets, he found a number of wreaths in the most perfect state of preservation imaginable.

" The roses, for instance," writes Mr. Newberry, " had evidently been picked in an unopened condition, so as to prevent the petals from falling. In drying in the coffin, the petals had shrivelled and shrunk up into a ball, and when moistened in warm water and opened, the androecium appears before the eye in a wonderful state of preservation. Not a stamen nor an anther is wanting—one might almost say that not a pollen-grain is missing. When taken from out of the sand and dust of the cemetery, the vegetable remains were very dry and brittle, and in that state it was quite impossible to examine them. They were therefore soaked in cold, lukewarm, or hot water (according to the species), when they soon recovered their original flexibility, and permitted of being handled and examined with ease. By this means it was possible to prepare a series of specimens gathered two thousand years ago, which are as satisfactory for the purposes of science as any collected at the present day."

More than twenty species are represented in these wreaths, including the Marjoram (*Origanum Marjorana* Linné), the Bay Laurel (*Laurus nobilis* Linné), the Myrtle (*Myrtus communis* Linné), the Linden (*Tilia europæa* Linné), the Mignonette (*Reseda ordorata* Linné), the Ivy, the Scarlet Poppy (*Papaver Rhœas* Linné), the berries of the Woody Nightshade (*Solanum Dulcamara* Linné), the white Egyptian Water-Lily (*Nymphæa Lotus* Hooker), an Immortelle (*Gnaphalium luteo-album* Linné), a Jasmine (*Jasminum Sambac* Linné), a Nubian Heliotrope (*Heliotropium nubicum* Linné), and *Narcissus Tazetta* Linné.

Very few of these seem to have been wild in Egypt. The Woody Nightshade, the berries of which were found threaded on strips of the leaf of the Date-palm, and which is mentioned by Pliny as "used in Egypt for chaplets," the Marjoram, and the Ivy probably came from Greece, *Narcissus Tazetta* from Palestine, and the Mignonette and the Jasmine from still farther east.

Among the food-plants found associated with these wreaths were Barley, Wheat, Cabbage, Lentil, Lupine, Bean, Chick Pea, Pea, Onion, Vine and Currant, Almond, Peach, Cherry, Pomegranate, Coriander, Walnut, Hazelnut, and Mulberry. The occurrence of the fruit of a Southwest-Asiatic Cornflower (*Centaurea depressa* Bieb) with the Wheat, supports De Candolle's suggestion that Wheat cultivation may have reached Egypt from Mesopotamia. Cabbage, we know from ancient authors, was highly esteemed as a vegetable. Although De Candolle considers Lentils to have been first cultivated in prehistoric times in the Levant, the discovery by Dr. Schweinfurth of a mess of pottage made of them, in a tomb of the Twelfth Dynasty, shows that their cultivation must have reached Egypt at a very early date. The Lupine (*Lupinus Termis* Forskal) is probably a native of Syria; the Chick Pea (*Cicer arietinum*

Linné) also of south-west Asia. The Cherry was probably introduced into the gardens of the Fayum from Greece, while the Mulberry, judging from mention of it in early hieroglyphics, must have come from its Asiatic home at an even earlier period.

The records of the cultivation of the Grape, and the making of wine in Egypt, date back five or six thousand years. They appear, for instance, on the tomb of Ptah-Hotep, who lived at Memphis about 4000 B.C. The variety known as a Currant may, however, be of Greek origin.

Considerable interest attaches to the few remains of the Pomegranate found at Hawara. They contained only four chambers instead of the six or eight now usually present. This is only one among several instances of cultivated plants of these ancient remains that have smaller and less improved fruits and seeds than those of to-day. It is interesting to note, however, that the weeds associated with them show no evidence of any such alteration during the lapse of the several thousand years in question. The Pomegranate, rarely mentioned in the earlier hieroglyphics, was generally cultivated in Egypt by the time of the Eighteenth Dynasty, about 1500 B.C. The Israelites in the Book of Numbers speak of Egypt as " a land of Figs, Vines, and Pomegranates "; but many cities and districts in Palestine itself were named after this tree. De Candolle doubts whether it was wild in Palestine, deriving it from Persia, in spite of the later Roman blunder in its name *Punica*, implying an African origin. The earliest mention of the Pomegranate in Egypt is in the inscription at Thebes on the tomb of the scribe Anna, who died under Thothmes I. The earliest representation of the plant is in a tomb at Tell-el-Amarna of the time of Amenophis IV. This would agree with the introduction of this species from Asia by the Hyksos or Shepherd Kings, about 2000 B.C. The ancient Egyptians are said to have made a fermented liquor from its fruit, and it is interesting to learn that they were also acquainted with the now universally recognised value of its rind as a vermifuge.

Of such vital importance to man has the cultivation of Wheat become that we are not surprised either at the innumerable varieties of the cultivated plant, while its wild ancestry is but little known, or at the myths associated with the origin of its cultivation. The Egyptians attributed its discovery to Isis or Osiris; the Greeks, to Ceres or Triptolemus; and the Chinese considered it the direct gift of heaven. In bricks of the pyramid of Dahschûr, supposed to date from 3359 B.C., Unger found numerous grains of a small-grained Wheat apparently identical with the *Triticum vulgare antiquorum* derived by Heer from the earliest lake-dwellings at Robenhausen, which probably date to earlier than 1000 B.C. The cultivation of some form of Wheat in Egypt goes back to before the invasion of the Shepherd Kings. The cultivation of the six-rowed Barley (*Hordeum hexastichon* Linné) is, perhaps, even more ancient.

It has been suggested that the Egyptians derived Wheat, Barley, and Flax from very ancient contact with Chaldæa. Mummy-clothes were always of linen and priests were compelled to wear linen garments, while a thread of Flax in one of the Dahschûr bricks carries the cultivation back to nearly 4000 B.C.

Spinning and weaving are represented in the Twelfth Dynasty pictures at Beni Hassan. *Linum humile* Miller, which was probably the plant grown, is now cultivated in Abyssinia, but neither for fibre nor for oil—it is grown merely as a famine food for the poor. The early Egyptian paintings show that their Flax was not reaped but uprooted, so that it was an annual, not a perennial form. As the cultivation of the plant in Chaldæa goes back to a pre-Babylonian antiquity, De Candolle suggested that it was brought thence by the early Egyptians, whom he considers were a proto-Semitic race migrating from Asia across the Isthmus of Suez.

Some ancient paintings and mummies, after the Twentieth

Dynasty, show the existence of the practice, still widely spread in the East, of dyeing the nails and parts of the skin orange with the juice of the leaves of Henna (*Lawsonia inermis* Lamarck), and some twigs of this plant were found at Hawara It is valued also for its fragrant flowers, and, though recorded as more or less wild from China, Java, India, Ceylon, Baluchistan, Persia, and Nubia, is of very doubtful origin.

Among other miscellaneous vegetable substances recorded are the twigs and berries of a Juniper (*Juniperus phœnicea* Linné) used in stuffing the mummies of crocodiles; coffins made sometimes of Pine (*Pinus Pinea* Linné), though usually of Sycamore-Fig—this pine-wood (as was the Cedar of Lebanon, used for panels on which portraits were painted) was probably imported from Syria; baskets and rope of the fibre of the Date-Palm; and a pair of cork soles which would seem to point to early commercial relations between Egypt and Spain.

The chief resin employed in Egyptian embalming was that of the Aleppo Pine (*Pinus halepensis* Aiton). That species, and perhaps also the Cedar of Lebanon, was cultivated in ancient times in Lower Egypt.

Of the other species represented on Egyptian monuments, besides the Pomegranate, the Grape-Vine, and the constantly recurring Date-Palm, the most interesting are the Papyrus, the Blue and White Water-lilies, the *Nelumbium*, and the various edible species of *Allium*.

So abundant was the Papyrus (*Cyperus Papyrus* Linné) in ancient times that it formed the hieroglyphical symbol for the region. Its starchy rhizome or underground stem was cooked for food by the poor. It furnished a much-valued charcoal. Its long flexible stems, 10 ft. or more in height, were made into baskets or coated with bitumen to form coracles, such as the " ark of bulrushes " in which Moses was abandoned. Its white pith, cut into longitudinal slices and arranged in crossing layers and polished with ivory burnishers, constituted the first true paper. Introduced into Sicily, where it still flourishes, by Hiero of Syracuse, it was entirely neglected after the invention of rag-paper. It still existed at Damietta in the time of Napoleon and it is still abundant above Khartum. Its stiff, erect, broom-like umbels of many rays appear often on the monuments, notably in the picture of a Hippopotamus hunt in the Mastaba of the Ti in Saqqarah.

It is said that the rhizome of the white-flowered *Nymphæa Lotus* Linné has been used for food in Egypt since the time of Menes; and that the seeds were also eaten, as they still are on the Upper Nile. The blossoms were used to decorate the banqueting-hall—Egyptian ladies carried them in their hands or twined them in their hair. One of the garlands covering the mummy of Rameses II is composed of them. At Memphis, the dentate leaves, four sepals, white petals, and poppy-like fruit appear in paintings that are as old as the Pyramids.

The entire-leaved blue-flowered *N. cærulea* Savigny, described as the Blue Lotus by Athenæus in the second century A.D., appears less frequently on the monuments, but has been found in garlands on mummies.

There is no doubt as to the occurrence of the beautiful *Nelumbium speciosum* Willdenow in ancient Egypt, though it no longer grows in that country. It is, perhaps, nowhere indigenous in Africa, though widely distributed in Asia from Japan and Siam to Astrakan. Herodotus called it the Rose Lily of the Nile and described the obconic receptacle in which the edible seeds are sunk as being " like a wasp's-nest." Theophrastus describes it under the name *Cyamos*, comparing the peltate leaves to a Thessalian hat and the fruit to the rose of a watering-pot. Strabo describes how in his time (*i.e.* about the Christian era) the people of Alexandria loved to make boating excursions to a neighbouring mass of this plant and to breakfast under the shady leaves. It is true, however, that most representations of this plant, such as the mosaic in the Museo Borbonico at Naples, in which a Nile landscape is represented by this

C

plant and some crocodiles, are of a late Greco-Roman date, and that the only remains of the plant itself are those of the same age found by Professor Petrie at Hawara.

There is not much evidence of the use of Shallots (*Allium ascalonicum* Linné) or Leeks (*A. Porrum* Linné) by the ancient Egyptians. Herodotus tells us that Garlic (*A. sativum* Linné) was given to the builders of the Great Pyramid, and he is the sole authority for its use in ancient Egypt. No such doubt attaches to the Onion (*Allium Cepa* Linné), however, De Candolle considering it a native of " a vast area of Western Asia, extending perhaps from Palestine to India." Not only does Herodotus speak of it in the same sentence as he does of Garlic, but Juvenal's satirical reference to the divine honours accorded to it must have had some foundation. Many monumental representations are almost certainly bunches of onions, and Professor Petrie found many of the bulbs among the offerings made to the dead at Hawara.

On the walls of the Temple of El-Dêr-el-bachri is represented the collection brought back by Queen Misaphris or Hatasau, of the Eighteenth Dynasty, from her naval expedition against Punt or Somaliland about 1500 B.C. In this were thirty-one living specimens in tubs of the " Incense-tree," which she planted at Thebes. This was probably the Frankincense or Luban tree of that area (*Boswellia thurifera* Carter), and is an interesting instance of the commencement in warfare of that trade in spices with the eastern horn of Africa that was to become later of vast importance.

At Beni Hassan we have pillars with their capitals modelled on the bud of the Lotus. At Philæ they are ornamented with alternating Lotus flowers and Date-palm leaves. At Edfoo we have the Palm leaves alone; and at Luxor we find angular flutings representing the Papyrus.

CHAPTER III

ALMOST all the plants found by Professor Petrie at Hawara
—not only those in the garlands but also the edible species
in the funeral offerings—are garden plants of foreign (*i.e.*
extra-Egyptian) origin. From this we learn that there must
have been an extensive system of ornamental and kitchen
gardening in ancient Egypt. In this connection there was
found at Thebes an Eighteenth Dynasty plan of a villa, or
enclosed estate. The villa was situated by the side of a river,
and had several buildings and a garden with rows of various
trees and mixed groups of smaller plants.

There is no question as to the existence of intercourse of
ancient date, and recurring through long years of time,
between the valleys of the Nile and the Euphrates. Such
intercourse, though at times the hostile relation of war, was
without doubt at other periods one of commercial inter-
change and rivalry. It is difficult to determine the relative
antiquity of the two civilisations, but from a very remote
date Babylonia was a land of fertility. The cultivation of
the land was assisted by irrigation, and the country was a
plain of Date-palms with cities adorned by gardens.

"The land of Assyria," writes Herodotus in the fifth
century B.C., "is but little watered by rain, and that
little nourishes the root of the corn; however, the stalk
grows up, and the grain comes to maturity by being
irrigated from the river, not, as in Egypt, by the river
overflowing the fields, but it is irrigated by the hand
and by engines. For the Babylonian territory, like
Egypt, is intersected by canals, and the largest of these
is navigable, stretching in the direction of the winter

sunrise; and it extends from the Euphrates to the Tigris, on which river stood the city of Nineveh. This is, of all lands with which we are acquainted, by far the best for the growth of corn; but it does not carry any show of producing trees of any kind, neither the fig, nor the vine, nor the olive; yet it is so fruitful in the production of corn that it constantly yields two hundredfold, and at its best even three hundredfold. The blades of wheat and barley grow there to full four fingers in breadth; and, though I know well to what height millet and sesame grow, I shall not mention it; for I am well assured that to those who have never been in the Babylonian land what has been said concerning its productions will appear incredible. They use no oil except that extracted from sesame. They have palm trees growing all over the plain; most of which bear fruit from which they make bread, wine and honey. These they cultivate as fig trees are cultivated, tying the fruit of that which the Greeks call the male palm about the trees which bear dates in order that the fly entering the date may ripen it, lest otherwise the fruit fall before maturity."

Berosus, writing in the time of Alexander the Great, declares that Wheat, Barley, and the Date-palm were all wild in that region, a statement that De Candolle admits to be highly probable. Pliny adds that the Wheat was cut twice, and after that afforded good fodder for sheep.

With so fertile a soil it is not surprising that the Babylonians were an agricultural people. We learn that from a very early period, even in their cities, the houses of the well-to-do always had gardens alongside them. Originally, perhaps, the garden was little more than a grove of Date-palms; but, later, herbs and vegetables were also grown, and (says Professor Sayce)

" as habits of luxury increased, exotic trees and shrubs were transplanted to it and flowers were cultivated for the sake of their scent."

As early as 1130 B.C. Tiglath-pileser I, of Assyria, records

that he had taken Cedars and other trees from lands that
he had conquered and had planted them in the gardens of
his own land, and had brought thither rare Vines that had
not previously existed in Assyria. We read of Sennacherib,
four centuries later, laying out a " paradise " or pleasure-
park by the side of his palace, planted with Cypresses and
other trees and fragrant plants, and watered by a lake.
In one of the bas-reliefs from Kouyunjik, of this date,
figured by Layard, such a garden is represented, with Date-
palms, Cypresses, and smaller plants, irrigated by canals
and, perhaps, supplied with roads and paths. A broken
corner of the carving represents one of the terraces, or
" hanging gardens," supported upon pillars, for which
Babylon was so celebrated.

> " In Babylonia, in fact," says Professor Sayce, " an
> estate was not considered complete without its garden,
> which almost invariably included a clump of palms.
> The date-palm was the staple of the country. It was
> almost the only tree which grew there, and it grew in
> marvellous abundance. Stem, leaves, and fruit were
> all alike turned to use. The columns and roofing-
> beams of the temples and houses were made of its stem,
> which was also employed for bonding the brick-walls of
> the cities. Its fibres were twisted into ropes, its leaves
> woven into baskets. The fruit it bore was utilised in
> many ways. Sometimes the dates were eaten fresh, at
> other times they were dried and exported to foreign
> lands; out of some of them wine was made, out of
> others a rich and luscious sugar. It was little wonder
> that the Babylonian regarded the palm as the best gift
> that Nature had bestowed upon him. Palm-land
> necessarily fetched a higher price than corn-land, and
> we may conclude, from a contract of the third year of
> Cyrus, that its valuation was seven and one-half times
> greater."

On the ancient Assyrian monuments we have the con-
ventional so-called " Greek Honeysuckle " pattern. This
may be based on a group of hooded cobras seen in full face

and in profile, or upon the buds of the Egyptian *Nymphæa Lotus*; rosettes resembling Daisies; and many representations of the Date-palm, the Grape-vine, the Fig, the Pomegranate, and some kind of Melon. Masses of the Reed *Arundo Donax* Linné are frequent in the bas-reliefs reproduced by Layard. Some sculptures seem to be unquestionable representations of *Lilium candidum* Linné, a native of Palestine and the Caucasus. Whether others represent the Banana, as sometimes has been claimed, is doubtful.

Whatever may have been the characters of the gardens said to have been laid out by Semiramis at the foot of Mt. Bagistanos and to have been visited by Alexander, it would seem that the love of parks and trees possessed the ancient kings of Persia as it had done those of Assyria. The flame-like form of the Cypress (*Cupressus sempervirens* Linné), believed to have been originally a native of the mountains west of Herat, seemed to fit it in the minds of Zoroastrians for the precincts of the Temple of Fire.

> "A slender Cypress," writes Firdusi in his *Shahnameh*, "reared in Paradise, did Zerdusht plant before the gate of the temple. . . . When its top was surrounded by many branches, he encompassed it with a palace of gold and published abroad the challenge:—Where is there on the earth a cypress like this of Kishmir?"

When, about A.D. 850, the Caliph Motewekkil cut down all the sacred Cypresses of the Magians, this one is said to have shown 1450 annual rings of growth.

The Sanskrit word *Pradêsa* or *Paradêsa*, meaning a district or a foreign land, may be connected with a Babylonian word found in the exercise-book of a Babylonian school-boy as the name of some mythical locality. It seems to have acquired in the Zend, or ancient Persian, the sense of a pleasure-park, and to have passed from Persian into the Hebrew *pardês*; translated as "garden" and "orchard" in the Books of Canticles and Ecclesiastes; into the Arabic *firdaus*, plural *faradisu*, in the Koran; and the Greek *paradisos*. We have

ceased to use it as a general name for a park or garden,
though this sense is for ever associated in the minds of all
lovers of quaint old books on plants with John Parkinson's
title, punning on his own name, *Paradisi in sole Paradisus
terrestris*, "Parkinson's Earthly Paradise."

Though there does not appear to be much to be said
ethnologically in favour of the theory that would derive the
Chinese from any Mesopotamian race, a love of park-like
gardens serves to link these two distant parts of the continent of Asia. The legendary founder of Chinese agriculture and medical botany is the Emperor Chin-nung
(2737 to 2697 B.C.). He is said to have invented the plough
and to have instituted the annual ceremonial sowing (by
the emperor) of the seeds of five pre-eminently valuable
plants. Rice must be sown by the emperor in person;
Wheat, Millet (*Saturia italica* Beauvois), Sorghum, and the
Soy-bean (*Glycine soja* Bentham) may be planted by the
princes of his family.

Chin-nung is said to have discovered seventy poisons and
their seventy antidotes in one day, and to have had a glass
front fitted to his stomach to observe the processes of
digestion! He also wrote a pharmacopœia, the nucleus of
the *Pun-tsao*, published some 4300 years later. His image
is found in every Chinese druggist's shop to this day.

Under the Han dynasties (206 B.C. to A.D. 9) gardens
are said to have become so extensive as to damage agriculture and excite the populace to revolt. It would be
difficult to improve upon the general principles of garden
design laid down by the ancient Chinese writer, Lieu-
tscheu, who tells us that:

> "The art of laying out gardens consists in an en-
> deavour to combine cheerfulness of aspect, luxuriance
> of growth, shade, solitude, and repose in such a manner
> that the senses may be deluded by an imitation of rural
> nature. Diversity, which is the main advantage of
> natural landscape, must, therefore, be sought by a
> judicious choice of soil, an alternation of chains of hills

and valleys, gorges, brooks, and lakes covered with water plants. Symmetry is wearisome, and ennui and disgust will soon be excited in a garden where every part betrays constraint and artificiality."

So isolated and so stationary has Chinese civilisation been through long ages that it has only been at considerable intervals, and by the borrowing of solitary plants, that it has exerted any influence on the world at large, until the present day.

The great pharmacopœia *Pun-tsao*, compiled by Li-Shi-Chin between 1560 and 1590 A.D. from 800 previous writers, containing nearly two thousand prescriptions and occupying some forty volumes, has contributed nothing of value to medical science. The descriptive poem on the Manchu capital of Mukden, written by the Emperor Kien-long in the eighteenth century, has been translated into a European language and is said to dwell with genuine appreciation on the characteristic forms of different species of trees, their branching and their foliage; but it is certainly not generally known.

Some plants of the greatest possible value for beauty and for use, however, have been obtained by Europeans from China. As early as 1689 there were six varieties of *Chrysanthemum sinense* Sabine in the gardens of Holland. Osbeck brought *Chrysanthemum indicum* L. from Macao in 1751, and Philip Miller had it in cultivation at Chelsea in 1764. Sir Joseph Banks re-introduced *C. sinense*, the chief parent of our cultivated Chrysanthemums, in 1789. Since then our flower-gardens have been enriched with many beautiful species, especially Primulas, Poppies, and Rambler Roses, brought home by various collectors. In 1840 Robert Fortune's introduction of the Chinese Tea-plant (*Thea-sinensis* Linné) into India was destined to revolutionise a world-wide industry.

Of the introduction of the Mulberry and the silkworm into Europe, and of the travels of Marco Polo, we shall speak later.

CHAPTER IV

THE PLANTS OF THE OLD TESTAMENT AND THE WISDOM OF SOLOMON

ALTHOUGH we do not look to Holy Scripture for scientific precision of language as to physical nature, it is somewhat remarkable that in addition to the mention of nearly 120 plants from first to last (*i.e.* from Genesis to the Apocalypse), we have in the very first chapter of Genesis a concise practical classification of the more familiar plants, and in the next chapter an equally simple summary of the ideal of a garden. Plants are simply divided into " grass, the herb yielding seed, and the tree yielding fruit " enclosing seed. Until the eighteenth century, botanists were content with a primary division of plants into herbs, shrubs, and trees, or into herbaceous and woody plants.

When we are told that " God planted a garden eastward in Eden," it is added that He made to grow out of the ground " every tree that is pleasant to the sight and good for food." The Oriental notion of a garden was primarily a plantation of trees, combining the æsthetic objects of pleasure to the sight, including the provision of shade, and, very generally also, of fragrance as well as colour, with the utilitarian provision of edible fruits.

Most of the plants mentioned in the Old Testament are those that were either natives of, or were commonly cultivated in, Palestine. There are also references to the edible plants of Egypt, to the Manna of the Desert, and to various perfumes and costly woods that were articles of commerce from distant lands. Many references, such as those to " thorns and thistles " and to " bitter herbs," are, of course, of a purely literary generality, unconnected with any par-

ticular species. Others are exceedingly difficult to identify
with certainty, and many have been associated by non-
botanical translators with European names of plants that
could never have occurred in Palestine.

The dry, warm and sunny climate of Palestine and its
mostly light calcareous soils tend especially to produce the
more xerophytic or drought-bearing type of vegetation. At
the same time they are favourable to the rapid development
of well-ripened crops of Wheat, Barley, and Grapes. The
stony waste ground produces a variety of spinous plants,
including such shrubs and trees as *Zizyphus spina-Christi*
Willd., and the Bull's-horn Acacia (*Acacia Seyal* Delile) and
many smaller plants. These, being of no particular value,
are merely referred to in general terms as indicative of
sterile land. Many of the native trees and shrubs are of the
evergreen, sclerophyllous, or leathery-leafed, and often
aromatic type so characteristic of the Mediterranean flora,
such as the Bay (*Laurus nobilis* Linné), the Myrtle (*Myrtys
communis* Linné), the Box (*Buxus longifolia* Boiss.), the Olive
(*Olea europæa* Linné), and the Fig (*Ficus Carica* Linné).
In the flora are also many herbaceous plants with pungent
or bitter juices, such as Mint (*Mentha sylvestris* Linné) and
Wormwood (*Artemisia*).

For convenience we may group the chief plants mentioned
in the Old Testament in the eight classes:—(i) timber and
shade trees; (ii) fruit-trees; (iii) cereals; (iv) pulse; (v)
bulbs, gourds, and other edible plants; (vi) flax; (vii)
perfumes; and (viii) a few miscellaneous plants that do not
come conveniently under the previous seven heads.

(i) The Gopher wood of Noah's Ark was probably the
Cypress (*Cupressus sempervirens* Linné) that is still abundant
in Armenia. Because English-grown Cedar (*Cedrus Libani*
Loudon) is not strong, and because the species occupies
but a small area on Lebanon to-day, it has been doubted
if this wood was used by Solomon in the Temple. The
discovery by Layard of a beam of it in one of the palaces
of Nineveh, however, shows that it was used for such

purposes. The Fir that is said to have been also supplied by Hiram of Tyre to Solomon may well have been *Pinus halepensis* Aiton, the Aleppo Pine, common to-day on the Lebanon range.

The Date-palm (*Phœnix dactylifera* Linné) was more abundant in the land than it is now. Its Greek name gave the name Phœnicia to the country, and its beauty caused its Hebrew name Tamar to be adopted as a woman's. The Palm-groves of Jericho, given by Antony to Cleopatra, were farmed for her by Herod the Great. Solomon's Temple seems to have been ornamented with alternating Date-palm leaves and "open flowers." probably Lotus, as are the capitals at Philæ. The abundance of this Palm about Jerusalem led to its use by the populace on the first Palm Sunday. From the absence of allusion to the fruit, it was perhaps then as little developed in Palestine as it is to-day in Greece and Italy.

The Shittim wood used in building the Tabernacle was no doubt the dense orange timber of *Acacia Seyal*, almost the only species at hand. Several passages suggest that the wood of the Sycamore Fig (*Ficus Sycomorus* Linné) was used for common carpentry in Palestine, as in Egypt. In the small fruit of this species, which is much inferior to that of the Common Fig (*Ficus Carica* Linné), it is necessary to cut off the top to allow the acrid juice to escape before it ripens; this operation was the humble task of the prophet Amos.

Among the rare woods, imported by Phœnician commerce from afar, was the black heartwood of the Ebony (*Diospyros Ebenum* Linné), mentioned by Ezekiel and brought apparently by coasting trade from Malabar, by way of the Persian Gulf. Included among these was also the Almug or Algum trees, brought from Ophir for Solomon by Hiram, of which were made pillars and musical instruments. This was probably the Red Sanderswood (*Pterocarpus santalinus* L. fil.) of India and Ceylon, a very hard dense wood susceptible of a high polish and used even to-day for carving and turnery.

The most common Oak in Palestine is the evergreen

Quercus pseudococcifera Desf., to which species the famous so-called Abraham's Oak at Hebron belongs. Not only has this tree replaced a more ancient Terebinth (*Pistacia Terebinthus* Linné), but the Hebrew name *elah*, meaning Terebinth, has often been translated by " Oak."

(ii) The first tree mentioned in the Bible is the Fig (*Ficus Carica* L.), a native of Palestine. The " Land of Promise " is described as a land of Vines, Figs, Pomegranates, and Olives. As in the case of the Oak, the Plane, and the Sycamore, the Fig and the Vine were valued for their shade as well as for their fruit, every man sitting under his own Vine or Fig tree representing the ideal of peaceful liberty. No doubt from very early times, as in the East to-day, the wholesome and nutritious dried fruit of the Fig formed an important part of the food of the people.

The antiquity of the manufacture of wine is indicated by its association in Scripture with the name of Noah. The " treading of the wine-press " figures in very ancient monuments. Sun-dried raisins may be of equal antiquity and are now relatively more important in Mohammedan countries, where wine is forbidden.

Though of comparatively late introduction into Egypt, the Pomegranate may well have been a native of Palestine, where its Hebrew name *Rimmon* occurs as a frequent place-name. The capitals of the pillars of Solomon's Temple were modelled on its fruit, which appeared also on the embroidery of the robe of the High Priest. The copious seeds seen through the bursting sides of the fruit seem to have suggested fecundity.

In a Mediterranean land of deficient rainfall, however, where grass is scarce and butter almost unknown, the Olive becomes the most valued of fruit trees. Its Oil is used as an illuminant, as an unguent, and as an article of food; it gives man " a cheerful countenance." The fruit-bearing trees require to be grafted; and the wood of the wild tree, with its remarkably streaky yellows and browns, is as ornamental as that of trees more valuable for other uses.

The ancient Olive trees of Gethsemane are far dearer for their associations to the Christian, whatever his nationality, than can be the most beautiful groves of the tree in Italy or elsewhere that serve merely to arouse feelings of patriotism or the artistic appreciation of form and colour.

The Almond (*Amygdalus communis* L.) is another of the native fruits of Palestine. The Pistachio-nut, if not also native, has long been largely cultivated there; and, since the tree (*Pistacia vera* L.) was less suited to the climate of Egypt, its nuts, with Almonds, made an appropriate present for Jacob to send to his son Joseph when the latter was Governor of Egypt. The " nuts " of the Song of Solomon were probably Walnuts, originally, perhaps, native to Western China but long cultivated further west. The " Apples of gold " were probably Apricots, an early introduction from Armenia.

(iii) Though inferior in food-value to Wheat, Barley was even better suited to the light soil of Palestine, and in addition required at least three weeks' less time to reach maturity. Wheat is still the general food for draught animals in Palestine; but Barley loaves, and even more so those made from the Millets, were rightly considered far inferior to those of Wheat as human food. The Land of Promise was " a land of Wheat and Barley "; and, from the time of Solomon, Palestine was a corn-exporting country. Wheat and Spelt (*Triticum Spelta* L.), called " rye " or " fitches " in our Authorised Version, are sown in Palestine in November and December and reaped in May.

(iv) We have already referred to the antiquity of Lentils, which were occasionally used as a bread-stuff, though usually as a porridge.

(v) Though Leeks, Onions, and Garlick are mentioned in the enumeration of good things in Egypt regretted after the Exodus, it is not improbable, judging from the monuments, that the last-named may have been the Shallot (*Allium ascalonicum* L.). Certainly they were all edible bulbs of species of *Allium*, and they may well have been missed in the desert, since they are particularly valued in the East as preventives of thirst, as also are Cucumbers and Melons, mentioned in the same connection. The common Cucumber (*Cucumis sativus* L.), the Melon (*C. Melo* L.) and the Water Melon (*Citrullus vulgaris* Schrad.) are all largely cultivated to-day both in Egypt and in Palestine; and, although the Castor-oil tree (*Ricinus communis* L.) has been very confidently identified with the *Kikaion* under which Jonah rested, the old rendering of the word as the Gourd (*Cucurbita Pepo* L.) is quite possibly correct.

As to Manna, it is now generally agreed that the small greyish lumps of the lichen *Lecanora esculenta*, which is often drifted in considerable quantities by wind in the desert regions of North Africa and South-west Asia, best answers the Biblical description.

(vi) Though it has been suggested that the pischtah, among which Rahab hid Joshua's spies, may have been Cotton branches, there is no satisfactory evidence for the cultivation of Cotton in Lower Egypt or Palestine before the Christian era. It remains to-day the custom in the East to spread out the stalks of Flax on the flat house-roofs to dry in the sun.

(vii) Much interest attaches to the various perfumes and spices mentioned in the Bible. With the exception of Saffron, the sun-dried stigmas of the Mediterranean species *Crocus sativus* L., they were none of them native plants.

The earliest glimpse we have of the spice trade is when the company of Ishmaelites or Midianites from Gilead, carrying spicery, Balm, and Myrrh down to Egypt, appear on the horizon at Shechem when Joseph's brethren had just cast him into a pit.

Balm of Gilead, the fragrant gum-resin of *Balsamodendron gileadense* Kunth or *Commiphora Opobalsamum* Engler, was cultivated in the fertile plain of Jericho. According to a later tradition, it had been originally brought to Solomon by the Queen of Sheba. Myrrh, the similar product of *Balsamodendron Myrrha* Kunth, used in the holy oil of the

Tabernacle and in embalming; and Frankincense or Gum Olibanum, the gum-resin of various species of the allied genus *Boswellia*, were also, no doubt, the products of trade—probably caravan trade—with Southern Arabia and Somaliland, their native country. The name *Regio Cinnamomifera* was applied by ancient geographers to Somaliland, but it is probable that the Cinnamon and Cassia (the inner bark of the Lauraceous *Cinnamomum zeylanicum* Breyne and *C. Cassia* Blume), ingredients in the precious ointment of the Tabernacle, both came from a much greater distance—perhaps by that coasting trade to which we have already referred. Aloes were no doubt the balsam of the North Indian *Aquilaria Agallochum* Roxburgh, known as Eaglewood or Lign-aloes. Spikenard was probably the dried spikes of the fragrant rhizomes of the Himalayan *Nardostachys jatamansi* D.C.

(viii) There are but few flowers mentioned either in the Old or the New Testament merely for beauty. The Rose of Sharon, a sweet-scented bulbous plant, is probably *Narcissus Tazetta* Linné, which we have mentioned among the garland-flowers of Egypt. Though *Lilium candidum* Linné, a cultivated plant from the north, may have been the *shosan* from which some of the pillars of Solomon's Temple were modelled, as we have seen, it appeared in Assyrian sculpture, and any reference to a widespread brilliantly-coloured flower, such as some passages suggest, is more likely to have referred to the scarlet *Anemone coronaria* Linné.

It is very difficult to identify the *dudaim* or love-apples of Genesis xxx., translated " Mandrakes." They may have been the disappointing yellow apple-like explosive fruits of *Calotropis procera* R. Br., filled as they are with fine hairs mixed with the seeds and even yet reputed as a powerful love-philtre. Rabbinical and mediæval commentators have associated the name with *Mandragora officinarum*, however, which also has a large yellow scented fruit, and of this we shall have more to say later.

It is also difficult to identify that insignificant plant the Hyssop, chosen by the chronicler as the antithesis to the glorious Cedar in describing the encyclopædic botanical knowledge of Solomon. Though we cannot seriously accept Carlyle's etymology of the king as the cunning or kenning man, the man who knows, there is certainly a dignity in the ancient notion that the sovereign was supreme in knowledge as in power. If, in other cases, it would seem that the chief use of their botanical knowledge was to provide these monarchs with antidotes to the many poisons with which they were assailed, the portrait of Solomon praying for " an understanding heart " that he may judge his people and " discern between good and bad," is certainly more dignified than that of Mithradates. Josephus adds to the statement in the Book of Kings that Solomon " was not unacquainted with any of the natures of plants, nor did he omit to make inquiries about them, but described them all like a philosopher, and demonstrated his exquisite knowledge of their several properties," As for the " Hyssop that groweth out of the wall," a mere bundle of twigs convenient for use as a sprinkler, it may well have been the much-branched Caper (*Capparis spinosa* L.), commonly to be seen sprouting in the crevices of the rocky walls in the desert.

In the Song of Solomon we have only an indirect picture of a royal garden, enclosed, shaded with fragrant Walnut trees and Apricots laden with golden apples. Here the Anemone pleased the eye, every variety of aromatic shrub ministered to the sense of smell, and the Vine and the Pomegranate yielded their fruit. It is remarkable, however, that in the great body of Jewish interpretative case-law known as the *Mishnah*, which was handed down orally from about the Christian era, the very first group of treaties deals with agriculture and horticulture, prescribing the division of crops in fields, and the setting apart of the produce of each seventh year for the poor.

In a later age, under the tolerance of the victorious Saracens, as the Nestorians found their way into Mohammedan households as tutors, so did the Jews as physicians. They were associated with the Nestorians in the translation of many Greek philosophical works into Syriac, whence they were retranslated into Arabic. Thus they were partly instrumental in turning the hitherto obscurantist Moslems into the patrons of learning—but this is an anticipation of our story by many centuries.

CHAPTER V

PHŒNICIAN COMMERCE AND GREEK MYTH

It is not easy to keep to a strictly chronological order in tracing the growth of man's knowledge of plants, nor do the parallel lines of history always impress upon the mind the contemporaneity of events. The fourth Egyptian dynasty, under which the first pyramids were built, has been fixed at about 4000 B.C. Phœnician (and probably also Indian and Chinese) civilisation is believed to have been already existent. The twelfth Egyptian dynasty and Sargon, king of Babylonia, have been placed in the twenty-eighth century B.C.; the Golden Age of ancient China in the twenty-fourth; the Hyksos or shepherd kings in Egypt, the life of Abraham, and the height of Babylonian civilisation about the twentieth or nineteenth. The Mycenæan age and the beginnings of Greek civilisation are dated from the sixteenth to the eleventh century B.C., and the Siege of Troy, Moses and the Exodus in the thirteenth, or twelfth, and Solomon, Hiram of Tyre, and Homer in the tenth. We are thus justified in dealing with Phœnician commerce and Greek myth before we speak of the Homeric poems.

The history of geography practically begins with Phœnician enterprise. Though the wonderful voyagers of these times had always a commercial object, they were practically the earliest explorers. It so happened that they were shut into a narrow strip of coast by the Lebanon range, which then bore an abundance of timber suitable for shipbuilding, and no doubt these two factors played a powerful part in their subsequent development.

The Phœnicians seem at first to have been fishermen—the name Sidon means " the fishers' town." Steering by the sun by day and the pole-star by night, they coasted their way from island to island—to Cyprus, Rhodes, Crete, Samos, Thasos, and the shores of the Morea—in search of the various species of Mollusca from which they extracted their celebrated purple dye. Their proficiency in the art of dyeing may have been derived from Babylonia, and that of glass-making, in which they excelled at an early date, from Egypt, where it seems to have been of a very great antiquity.

Very probably their stations for fishing for the purple-yielding shell-fish may have been their earliest settlements outside Phœnicia. They borrowed a system of weights and measures from Babylonia. To them perhaps we owe the first stamped coinage and almost certainly the earliest alphabetic writing. Their glass, their choice woven and dyed stuffs, and probably spices and other Indian produce, brought by overland routes of unknown antiquity, provided them with ample stock for their widespread activities in bartering. After traversing wide expanses of the deserts of south-west Asia, these ancient overland routes utilised the sweet waters of the oasis of Palmyra and Damascus; and by them the Phœnicians may have received the Frankincense of south-west Arabia, and possibly gold from Mashonaland, ivory and Cinnamon from India. If the Semitic name kinnamon is made up of mon, meaning " sweet " (as in manna, or maun, the Tamil for a tree), and kin, the name for China or Siam, it may well be that the source of the much-appreciated spicy bark was quite unknown in early times.

From quarrying the marble of Paros and mining the copper of Cyprus and the gold of Thasos, the Phœnicians, by the twelfth century B.C., had reached the tunny-fisheries of the western Mediterranean, and even of the Atlantic. They had established themselves in Sicily and, perhaps, in Corsica, Sardinia, and the Balearic Islands, and were working the copper, silver, and tin of Tarshish.[1] The mere mention of Tarshish in the tenth chapter of Genesis suggests the pro-digious antiquity of some commercial knowledge of the

[1] Southern Spain.

farthest extremity of the Mediterranean. Not only did
Corinth, already celebrated for its wealth by the time of
the *Iliad*, bear the purple-mussel on its coinage; but the
Phœnicians appear to have been at Marseilles before the
arrival of the Phocæans in 600 B.C. They seem to have
established themselves at Cadiz (Gades, Gadeira, or Agadir)
before 1100 B.C., and at Utica on the north coast of Africa
a little later. (Carthage was not founded till nearly three
hundred years later.)

The Cypress seems to have been brought westward from
its home in Afghanistan, though doubtless derived from
Chaldæa, before the period of the *Iliad*. The Pomegranate,
mentioned in the *Odyssey* but not in the *Iliad*, was introduced
so early into the island of Cyprus as to have a divine origin
ascribed to it. It may well have been to the Phœnicians
that Delos owed its first Date-palm, mentioned both in the
Odyssey and in Euripides' *Hecuba*. This palm, which may
have given its name *Phœnix*, to Phœnicia, does not ripen its
fruit in Greece.

We are told that it was from Ezion Gebir and Elath,
which were at the head of the Gulf of Akabah, that Solomon's fleet, manned by Hiram's Phœnicians, started for
Ophir. The double journey involved a three-years' voyage,
and their return with a cargo of gold, silver, Algum-wood,
precious stones, ivory, apes, and peacocks, points almost
certainly to their having reached the Malabar coast. Algum
is said to be the Sanskrit *valgu ;* the Hebrew *shen-habbim* for
" ivory," to be connected with the Sanskrit *ibha*, " the
elephant." The Homeric Greek for ivory, *elephas*, was
this word with the Semitic article, i.e. *el-ibha*. The Hebrew
koph, for " apes," has no Semitic analogue, but is the Sanskrit
kaſi. *Tokki-im*, for " peacocks," represents the name *togei*,
which continues in use to-day on the Malabar coast. Though
the voyage may be apocryphal, the mere suggestion that,
at the command of Pharaoh Necho of Egypt (about 600 B.C.),
the Phœnicians in three years circumnavigated Africa, starting by the Red Sea and returning by way of the Straits of

Gibraltar, shows their well-recognised position as being the greatest navigators of the age.

That we know so little as to the discoveries of the Phœnicians is largely due to the entire loss of the literature of Phœnicia and of Carthage. It is also partly due to the trade jealousy that was the probable origin of all the wonderful travellers' tales of the difficulties in obtaining foreign products. These first appeared in Herodotus, and so passed into the literature of marvels.

> "The Arabians," we are told, "gather frankincense by burning styrax . . . for various small winged serpents guard the trees that bear frankincense, a great number round each tree, and are driven away by nothing but the smoke of the styrax. . . . Cassia grows in a shallow lake, around which lodge winged animals, very like bats, exceedingly fierce. These the Arabians keep off by covering themselves with hides, and so gather the cassia. Cinnamon they collect in a still more wonderful manner. In what land it grows they cannot tell, though some say that it is in those countries where Dionysius was nursed. Large birds, they say, bring the rolls of bark we call cinnamon for their nests, which are built with clay against inaccessible precipices. The Arabians, having cut up in large pieces the carcases of oxen, asses, etc., lay them near the nests and retire to a distance; and the birds carry these pieces of meat up to their nests, which not being strong enough to support the weight, break and fall to the ground, when the men come up and gather the cinnamon.

Such fables have their modern analogies in the haunted houses where ghosts conveniently protected the operations of smugglers. They were probably also circulated by Greek and Chinese drug-collectors, who wished to enhance the value of their wares.

In spite of Phœnician secrecy, however, their trade enterprise was soon followed by the Greeks, especially by those of the Ionian colonies on the Asiatic shore, immediately north of Phœnician territory, where a much-indented coastline provided numerous sheltered bays, with steep easily-fortified heights close at hand.   To the trading spirit of the Phœnician, however, the Greek added a spirit of philosophical inquiry, and so laid the foundations of the modern European sciences—geography, astronomy, and—later on—biology, psychology, and sociology.

As to the date of the contact between Phœnicia and Greece, it is noticeable that Phœnicia is only once mentioned in the *Iliad*, though often in the *Odyssey*, while the name of Tyre does not occur in either poem.

The many charmingly poetical fancies that the Greeks associated with their native plants belong to an earlier date than the Greek colonies.   Daphne, daughter of the river Peneus, pursued by Apollo, the Sun-god, was turned into the Laurel that grows by her paternal stream.   Persephone, the seed-corn, daughter of Demeter, goddess of fertility, carried off to the under-world from the flowery fields of Enna in Sicily, is allowed to return for two-thirds of the year, the period from sprouting to harvest, when the earth rejoices. The beautiful youth Hyacinthus—beloved by both Apollo, the Sun-god, and Zephyrus, the west wind—was killed by Apollo's quoit blown aside by Zephyrus, and from his blood springs a flower with $Aἰαῖ$, " alas," in dots on its petals.

The Phœnician personification of the annual revival and death of Nature was Adonis, from whose life-blood sprang the scarlet Anemone in the spring.   He fills the world with grief for his death during half the year, but is permitted to come up from the nether world to spend the other six months with Aphrodite or Astarte, the Syrian goddess of love and beauty, whose cult reached Greece by way of Cyprus.   She it was who planted the Oriental Pomegranate —emblem of fecundity—in Cyprus, as Athene gave the Olive to Athens.

As the Chariot of the Sun, driven by " the shining one " Phæthon, fell westward towards the dimly-known waters of the Eridanus, the Po, the sighing Poplars on its banks represented his sisters, the Heliades who had yoked the horses. Their tears became the amber, the *electron* or sun-stone which, in fact, only reached that region by the southern overland route from the shores of the Baltic. Pliny says that it was Mithradates who first correctly pointed out that amber was the resin not of a Poplar but of a Pine.

The wild men of the Thessalian forests, who hunted the wild ox on horseback, became to the Greek imagination the Centaurs, half man, half horse, just as did the Spanish cavalry when first seen by the natives of America. Chiron, son of Cronos, god of age, and Philyra, who became a Linden tree, was taught by Apollo and Artemis, by sun and moon, as every woodman is. Thus he became most skilled in bodily exercises, in hunting, in music, and in the healing art. Among his pupils were Æsculapius, the " blameless physician " of Homer (father of Podalirius and Machaon, the physicians of the Greek army before Troy), who was recognised in later times as god of medicine; Achilles, and Hercules. Hercules it was who accidentally caused the death of his master with a poisoned arrow. *Centaurea, Chironia, Achillea, Heracleum,* and *Asclepias,* among our plant-names preserve to this day the memory of this legendary school of wood-craft.

CHAPTER VI

THE PLANTS OF HOMER

ABSURD as it may seem, Strabo, in the first century, treats Homer as the founder of scientific geography and believes him to be practically infallible. It is abundantly clear that the topographical knowledge of the author of the *Iliad* was restricted to a comparatively small area, and that of the author or, as Samuel Butler maintained, authoress, of the *Odyssey*, was not much more extended. Any detailed knowledge on the part of the author of either poem is confined to the northern Ægean and the lands immediately contiguous to it. There is no allusion to Assyria, and neither the Euxine, the Bosphorus, nor the Scythians are mentioned. In the *Iliad*, Phœnicia is only once mentioned; Egypt only indirectly; and the " blameless Ethiopians," or " burnt-faced ones," are scorched by living near the ocean and the sun! The Ægean is *the* sea, as the Mediterranean was to the Greeks of a later date, and the Achelous (Aspro-potamo) was the greatest known river.

Within his limited area, however, Homer shows by casual epithets his undoubtedly personal knowledge. The characteristic grikes of a limestone area are recognised in " Lacedæmon, full of fissures." The Cyclopean fortifications of Tiryns are summed up in the one word " well-walled." Thisbe in Bœotia is aptly described, as it was in the time of Strabo and is to this day, as " full of wood-pigeons."

So too, with reference to the plants mentioned, it has been pointed out that those in the *Iliad* well represent the xerophytic maquis vegetation of the Troad. We have the Tamarisk; the Myrtle; the Box; the Manna Ash (*Fraxinus*

46

Ornus L.); the Olive; a Rose (*Rosa gallica*, L. or *R. semper-virens* L.); Pines, apparently including the Aleppo Pine (*Pinus halepensis*, Mill.), and, perhaps, *P. austriaca* Link. Some Oaks, probably evergreen species (as *Quercus Ilex* L. or *Q. pseudococcifera* Desf.); the Chestnut (*Castanea sativa* Mill.), with its edible fruit sacred to Jove; the Cypress; the Plane; the Saffron Crocus (*Crocus sativus* L.), and the Narcissus. The Cypress is only referred to indirectly as a place-name, and there is but one reference to the Plane (*Platanus orientalis* L.), an interesting species in that it apparently migrated from the East by way of the north of Asia Minor and not by the Semitic lands to the south.

Swampy river-sides are represented by reeds and sedges, used for thatch (probably *Arundo Donax* L., *Phragmites communis* Trin., and *Cyperus longus* L.). Among cultivated plants we have reference to the Almonds of Lemnos; the Fig; the Olive, and the Vine. The Onion is esteemed as a relish, and Spelt and Barley are apparently used as food for horses.

The *Odyssey* contains geographical allusions that suggest a somewhat later date, if not also a distinct authorship. Egypt and Phœnicia are mentioned frequently, and a voyage from Phœnicia to Libya appears as an ordinary occurrence. The Nile is mentioned as " the river Ægyptus," and there is the undoubtedly hearsay exaggeration of the dangerous Straits of Messina into Scylla and Charybdis.

A considerable number of plants is mentioned in addition to those of the *Iliad*. Here Ulysses refers to the Date-palm of Delos, and we have a solitary reference to the Bay Laurel. The " Cedar " used for fuel is probably *Juniperus Oxycedrys* L., and we meet with the black heart of Oak, no doubt one of the evergreen species, Oleander; Poplars, and what are apparently Alder and Willow. The rope made of " Byblos " may have been made of Papyrus. Even more problematical are Moly, Nepenthes, and the Lotus of the Lotophagi. Moly, Ulysses' preventive of intoxication, seems to be a species of *Allium*; but the statement that it had a yellow flower—as had Linnæus' species *Allium Moly*—rests only on Pliny's authority. Nepenthes is generally taken to be the Opium Poppy (*Papaver somniferum* L.), but Burton in his *Anatomy of Melancholy* insisted that it was Borage. Certainly the properties of Polydamna's gift to Helen were potent enough for anything. It frees men from grief and anger and causes oblivion of all ills. He who drinks it would not shed a tear for a whole day, even if his mother and father should die, or his brother or son be slain before his eyes!

The storm off Cape Malea (Capo St. Angelo) might well have driven Ulysses to the Syrtis Minor (the Gulf of Cabes and the region of the modern Tripoli). Thus *Zizyphus Lotus* L. was most probably the " honey-sweet fruit " that made those who eat it lose all wish ever to leave that land of the Lotus-eaters. Polybius, in the second century B.C., himself visited that part of Libya and gives the following account of the plant:

> " The lotus is not a large tree; but it is rough and thorny, and has a green leaf like *Rhamnus* but a little longer and broader. The fruit is like white myrtle-berries; but becomes purple and in size is about equal to that of the round Olive, and has a very small stone. When ripe it is gathered; and some of it is pounded up with the grain of Spelt and stored in vessels for the slaves, and the rest, after the stones have been taken out, is preserved as food for the free inhabitants. It tastes like a fig or a date, but has an aroma superior to that of either. A wine is also made of it by steeping it in water and crushing it, which is sweet and pleasant to the taste, like good mead."

It is in the *Odyssey*, too, that we have the description of the vineyard of Laertes and the still more detailed account of the garden of Alcinous in Phæacia, traditionally identified with Corfu. This latter garden was four acres in extent, surrounded by a hedge and watered by two streams, one of which seems to have been divided into irrigating ditches. In the vineyard portion some grapes were dried in the sun, others were trodden in the winepress. The fruit trees included also Pears, Pomegranates, Figs, Olives, and Quinces, the golden apples of which last-mentioned species (*Cydonia vulgaris* Persoon) gave its name to the island of Melos and appear on its early coins. There were also beds of various plants so chosen as to be flourishing all the year round. Bacon, it will be remembered, says:

> " I do hold it, in the royal ordering of gardens, there ought to be gardens for all the months of the year, in which, severally, things of beauty may be then in season."

In lieu of Alcinous's modest four acres, however, the great Elizabethan considered that a princely garden " ought not well to be under thirty."

CHAPTER VII

THE FATHER OF MEDICINE AND THE EARLY GREEK PHILOSOPHERS

During the century and a half between 740 and 600 B.C. the Greeks spread themselves along the shores of the Mediterranean and the Euxine. They followed up the Phœnician traders with a truer instinct for colonisation—much as in modern times the British have followed up the selfishly commercial Dutch settlements. Among their chief early objects was timber, which they required for ship-building and other purposes, and of which there was an inadequate supply in Greece. There were also the tunny-fisheries and the precious metals and wool to attract them to the new land, and the fertile volcanic soils of Sicily and Magna Græcia (*i.e.* Southern Italy) attracted them for the cultivation of the Olive and the Grape.

In the commercial explorations of the eighth and seventh centuries, and in the theoretical geographical science that arose from them in the sixth century B.C., the lead was taken by the Ionians of Miletus, on the coast of Asia Minor. Trapezus, now Trebizond, on the Black Sea, may have been the sea terminus of a prehistoric caravan route before they settled there. In the seventh and sixth centuries B.C. their settlements opened up trade with the Dnieper, the Danube and the Sea of Azov, as well as with Naucratis on the Canopic or Rosetta, mouth of the Nile.

About 630 B.C. the Dorians of Thera (Santorin) founded the exceptionally inland colony of Cyrene, round a spring in the centre of high ground that has been described as one of the most delightful and fertile places in all Africa. Here a brisk trade sprang up in the much-vaunted drug provided by the gum-resin of Silphium (*Ferula tingitana* L.). This is described by Pliny as "Hammoniacum," and he seems to have believed it to be distilled naturally in the sand with which it was mixed. This umbelliferous plant, which continues to be a striking object in the sands of Barca, appeared on the coins of the colony.

The foundation of Massilia (Marseilles) as a colony from Phocæa, a rocky promontory to the north of Smyrna, in 600 B.C., had most important consequences in opening up trade by way of the Rhone with Gaul and even Britain.

We are not much concerned with the early speculative philosophers of Greece. When a naturalist, trained in observation, draws generalisations from his own observations and from those of others, his conclusions may be of great value, based, as they are, upon facts. But when a philosopher condescends to mention plants or plant-life, his views are too often based rather upon what the facts ought to be, according to his preconceived opinions, than upon what they are. We have only fragments of the poem on Nature written by Empedocles of Agrigentum between 490 and 430 B.C. He seems to have resolved all things into the four elements, fire, air, earth, and water, and to have concluded that, as plants, animals, and man himself are all compounded of the same elements, there is a certain identity of nature in all of them. Democritus of Abdera, of whose writings also fragments only remain, interests us rather in that he is said to have been one of the teachers of Hippocrates, than for his theory of atoms. We do not know on what authority he is represented on the frontispiece of Burton's *Anatomy of Melancholy* seated in a walled garden, except that he is said to have written a book on the wonderful and magical properties of herbs.

Of greater importance to us are the writings of Herodotus of Halicarnassus in Asia Minor, which have come down to us intact. Born in 484 B.C., he travelled extensively in Egypt and Assyria as well as in Greece and Italy, and died about 424 B.C. Although the main purpose of his work is historical, it is full of geographical information and contains several most interesting references to plants. We have already quoted his version of the fabulous tales told to enhance the value of Arabian merchandise. In a sentence immediately preceding this account we have his description of Indian Cotton.

> " Certain wild trees there," he writes, " bear wool instead of fruit, that in beauty and quality excels that of the sheep; and the Indians make their clothing from these trees."

Elsewhere he speaks of Indian boats made of reeds, one joint making a boat, which would seem to refer to a giant bamboo.

It has been suggested that as early as the time of Homer there was not only a recognised medical profession but even some differentiation between the surgeon and the physician. Of the two sons of Æsculapius, Machaon was deputed to heal injuries, Podalirius to recognise what was not visible to the eye. Æsculapius himself seems probably to have been a native of Epidaurus—at all events, his chief shrine when deified was there, though others of importance were located at Cos and at Cnidos. His reputed descendants became a priestly caste, the Asclepiadæ, and among them the knowledge of medicine was regarded as a body of sacred secrets transmitted from father to son. To the Asclepia, or temples of Æsculapius, there seem to have been attached carefully conducted hospitals for the sick. On the columns or walls of the temple were inscribed votive tablets setting forth the symptoms, treatment, and results of each cure. From these the Asclepiadæ accumulated the clinical experience of their profession. Thus it came about that Hippocratic medicine was based rather upon observation than on theory.

Hippocrates of Cos, justly named " the Father of Medicine," is said to have been seventeenth or nineteenth in direct descent from Æsculapius, and to have been born in

460 B.C., dying nearly a hundred years later. There were no less than seven members of his family who bore the same name, however, and of the eighty-seven works ascribed to him, probably not more than seventy are his. Their composition was probably spread over a period of two hundred and fifty years, from the middle of the fifth century to the beginning of the third century B.C.

To Hippocrates is ascribed the merit of first insisting upon the natural origin of all disease and the necessity of tracing its natural history. To his school we owe the theory of the four humours in the human body, upon the due proportion—or " tempering "—of which health was supposed to depend. To these four humours—blood, phlegm, yellow bile, and black bile—our popular language is indebted for such words as sanguine, phlegmatic, bilious, and melancholy, and popular notions of physiology have been influenced by them even down to our own time.

Although here and there in the Hippocratic works some two hundred and forty plants are mentioned, it is only as drugs, with no attempt at botanical descriptions. As in the case of Homer, it is more easy to identify these plants by the use of their Greek names in later writers than it is to determine the identity of those mentioned by Hebrew names in the Old Testament. We can only mention a few—the red dye Madder (*Rubia tinctorum* L.); Scammony (*Convolvulus Scammonia* L.); Henbane (*Hyoscyamus niger* L.); Mandrake (*Mandragora Officinarum* L.); Hemlock (*Conium maculatum* L.); Chamomile (*Matricaria Chamomilla* L.), and the astringent rind of the Pomegranate, all became well known in medicine. The ranunculaceous *Helleborus orientalis* L. was curiously united with the liliaceous *Veratrum album* L. as Black and White Hellebore. The mention of Cardamoms (*Elettaria Cardamomum* Maton), a spice probably unknown to the Greeks until the invasion of India by Alexander the Great, stamps the particular treatise in which it occurs as belonging to the later Hippocratic school.

From very early times there existed in ancient Greece

men who made a business of collecting medicinal roots and herbs. They supposed the properties of a perennial to be concentrated in the underground parts, and because they attached most importance to the root they became known as *Rhizotomoi*, " root-cutters." Mingling magical incantations with their work, they are mostly spoken of contemptuously by Theophrastus and other men of science. Two or three of them seem to have been men of more culture, however, and even to have written books, none of which has survived. Cleidemus, for instance, is stated to have investigated the diseases of the Fig, Olive, and Vine and to have maintained that the organs of plants and of animals are analogous. Hippon argued that cultivated plants are derived from wild plants, and are liable to revert to their original condition.

CHAPTER VIII

IN several respects the marvellous military expedition of
Alexander the Great forms an epoch in geographical no less
than in civil history. In twelve years he not only sub-
jugated an area of country equal to that of Europe, but
fully two-thirds of that area was practically unknown to the
Western world until his time. In no one age—with the
sole exception of that of the discovery of America, eighteen
and a half centuries later—has one race come into contact
with more new aspects of nature or a greater mass of new
materials for the study of natural history. Full as are the
accounts that we have of Alexander's expedition, however,
they were not compiled until three or four centuries later.
Its results were made known mainly by the writings of
Strabo and Arrian, and it was through them that the Greeks
first became acquainted with the fan-like leaves of *Borassus
flabellifer* L., and the palm-wine made from it and other
species; the Banana; the Banyan; the cultivation of Rice
by irrigation, and the fermentation of its grain; the oil
of *Sesamum indicum* L.; the attar of Roses; the sugar of the
Sugar-cane, and the manufacture of paper from Cotton;
but these observations were not to be published by Aristotle
or even by Theophrastus.

It was no doubt the forests of *Cedrus Deodara* Loudon in
the Western Himalayas that supplied the materials for the
fleet of Nearchus; but of this there is no contemporary
record. It has been said, however, that about a quarter
of the plants described by Theophrastus, Dioscorides, Pliny,

E

and Galen were unknown in Europe before the expedition of Alexander.

It is undoubtedly by his marvellous contributions to mental and moral science that Aristotle (Plate I*a*), the great tutor of Alexander, ranks among the greatest names in the history of mankind. The scientific basis that he gave to comparative anatomy, however, well entitles him to the name of the Father of Natural History. He was born (probably in 384 B.C.) at Stagira on the Strymonic Gulf, some seventy miles east of Pella, the capital of Macedonia. It is noteworthy that he was of Asclepiad race and the son of a royal physician, which gave cause to Epicurus and other rivals to jeer at him as a drug-monger after his death. At the age of seventeen he went to Athens to study under Plato, and there he remained till 347 B.C. At Philip's invitation he became (in 342) tutor to the young Alexander, then fourteen years of age, and retained that post until Philip's death in 336. Aristotle then went back to Athens, where the Lyceum gymnasium, assigned to him by the State, became the centre of what was afterwards known as the Peripatetic school. Most of his works are believed to belong to the years between 335 and 323. After Alexander's death in the latter year, Aristotle was unpopular at Athens on account of his Macedonian sympathies and was charged with impiety. He went into voluntary exile at Chalcis in Eubœa, where he died a year after his illustrious pupil.

Aristotle was made known to mediæval Europe mainly by the translation and commentary of Averroes, and he was followed implicitly until the sixteenth century. As Dante says:

> " the master of the sapient throng,
> Seated amid the philosophic train.
> Him all admire, all pay him reverence due."

Although Galileo and Francis Bacon attacked him later, for making use of theoretical deduction rather than practical experiment, his History of Animals, at least, was based largely upon his own actual observations. On that ground,

(A) ARISTOTLE (384-322 B.C.)

(B) PLATO (? 426-347 B.C.)

facing p. 56.

zoologists from Gesner and Rondelet to Ray, Willughby, and Artedi have set a considerable value upon his zoological conclusions.

Aristotle points out that we can trace an unbroken chain from the lowest form of life to the highest, each group being divided from the next only by slight differences. As we cannot even say where plants and animals begin, since the " zoophyta " forms are so similar to both, we cannot tell in which division to place them. In the dictum that " Nature, in order to spend on one side, economises on another," we have the enunciation of that principle of " economy of nutrition " or " balancement of growth " that underlies so much of the laws of correlation.

Aristotle also argues that most plant life is much lower and less centralised than that of animals, since the parts of a plant are comparatively so simple and independent that if separated they will grow, whilst severed parts of animals die, and injury to one main organ causes the death of the whole body.

We might have expected Aristotle to have given us a treatise on plants to complete his survey of Nature, and as a matter of fact he more than once alludes to some works of his on the " theory of plants." Such works are specifically referred to as existing by Diogenes Laertius, Athenæus, and Nicander. At the same time, even if the small work that long passed as his is not so, we have many passages in his zoological treatises that set forth his views as to plants.

Plants, he says, have only the lowest form of life, self-nutrition. They do not move from place to place, but only exhibit movements due to growth or decay. They have no sensory faculty, although they are affected by some external influences, such as heat and cold. Plants feed by their roots as animals feed by their mouths. Their food must be liquid, but earth is combined in the water they absorb. Their food being already digested, the earth and its heat serving them as a stomach, they have no waste products. They do not sleep or wake, having no organs of sense, though they are affected by something like sleep. They have no distinction of sex, the male and female principles being so blended that they generate from themselves, producing seed from their superfluous food. Some plants do show some small difference of sex, however, as, for example, the Fig, an insect called *psen* flying out of the fruit of the wild Fig and entering the unripe fruit of the cultivated form and thus preventing it from falling off prematurely. Herodotus had recorded that for this reason growers plant branches or plants of the wild species near the cultivated ones. The sole duty of plants is to bear fruit and seed, but Willows and Poplars bear none.

Many of these conclusions are inaccurate, of course, but they evince a philosophical attitude of mind and considerable observation. Aristotle knew nothing of the action of chlorophyll, nor did anyone else until the time of Malpighi, in the latter part of the seventeenth century. That plants have no power of selecting nutritive matter from the soil and that they do, therefore, eliminate waste by-products of nutrition was first shown by Joachim Jung, whose works were not published until after his death in 1657. Predecessors of Aristotle—such as Anaxagoras, Diogenes, and Democritus—had held on theoretical grounds that all animals, even if destitute of lungs, must breathe. Anaxagoras is said to have argued that even plants had breath, but it was not until the work of Dutrochet in 1837 that plant respiration was definitely established. It is clear that Aristotle had recognised diœcism, or sexes on different individuals, but he did not recognise that " sterile," *i.e.* fruitless plants, as in Willows and Poplars, were male. He knew nothing of the functions of stamens and stigmas— nor, in fact, did anyone until the publication of the *Anatomy of Plants* by Nehemiah Grew in 1682.

As to the " caprification " of Figs, as it is called, Aristotle's observation was nearly correct. The insect in question is a small wasp now known as *Blastophaga grossorum*. The winged

female wasps leave the cavity of the wild Fig, and entering
that of a cultivated Fig, lay their eggs in the ovules or
immature seeds of some of the young fruits. From these
eggs are hatched wingless male wasps, and although they
do not leave the fruit in which they are bred, the females
carry pollen from one Fig to another. The insect cannot
complete its life-history in the cultivated Fig. Some Figs
in other parts of the world ripen without caprification, and,
indeed, many authorities now maintain that this primeval
practice is unnecessary. On the other hand, it is said that
the perfect Smyrna Fig can only be so produced. Authorities also differ as to whether cultivated Figs bear only female
flowers and are thus dependent on the *Blastophaga* for
pollination; whether it is at all necessary for the ripening
of the Fig that pollination should take place; or whether
the function of the insect is to stimulate development by
biting or burrowing into the Fig, as is the case with gall-flies.

From Aristotle's treatise on plants Athenæus quotes references to Dates without stones and grafted Pears. The
absence of any such passages in the work that for centuries
passed as Aristotle's is one of the chief arguments against
its authenticity. The work long known as Aristotle's *De
Plantis* exists in manuscripts at Basel and Wolfenbüttel and
was first printed in the *Geoponica*, attributed to the Emperor
Constantine VII, at Basel in 1539. Scaliger (in 1566)
disputed its authenticity; Mirbel (in 1815) called it—we
think unfairly—" a crude collection of mistakes and ab-
surdities " by an impostor. Ernst H. F. Meyer (in 1841)
argued very forcibly that it was the work of Nicholas of
Damascus, a Peripatetic philosopher and historian of the
first century B.C., and an intimate friend both of Augustus
and of Herod the Great. The work, as we have it, is
professedly translated from Greek to Latin, from Latin to
Arabic (apparently by Isaac Ben Honain, a Nestorian scholar
at the close of the ninth century), and from Arabic to
Latin by some mediæval scholar of the name of Alfred.
It is Aristotelian in its views and, to a great extent, in style,

repeating many of the conclusions in the other works of
Aristotle, as we might have expected Aristotle to do. It
formed the basis of the botanical work of Albertus Magnus
in the thirteenth century, he, of course, believing it to be
the genuine work of the Stagirite, as did all scholars until
the time of Scaliger. It divides plants into trees, shrubs,
grasses, and garden plants, or into house, garden, and wild
plants; discusses roots, bark, leaves, flowers, and fruits,
and contains allusions to the milky juices of some plants
and to the odoriferous plants of Syria and Arabia.

Theophrastus of Eresus, another and a favourite pupil of
the great philosopher, is the author of the earliest scientific
works on plants and is known as the Father of Botany. He
was born at Eresus in Lesbos (Mitylene), about 370 B.C.,
and had studied under Plato before he became the pupil
of Aristotle. Aristotle at his death bequeathed to him his
library and manuscripts and nominated him his successor
as head of the Lyceum. He is said to have had two
thousand disciples and to have composed more than two
hundred treatises, of which his two botanical works are the
most important that have survived. At his death (about
287 B.C.) Theophrastus bequeathed his house and garden—
in which seven slaves seem to have been employed—to ten
of his pupils for the use of the school. His manuscripts,
with those of Aristotle, he left to Neleus, one of the ten.

There is nothing to show that the garden was a botanical
one; but the story of the books is somewhat romantic.
Neleus returned to his home at Scepsis in the Troad, taking
them with him, and, according to Athenæus, he afterwards
sold them to Ptolemy Philadelphus, who transferred them
to Alexandria. Strabo, however, relates that many of them
remained in the possession of the heirs of Neleus, who concealed them in a vault to escape the requisitions of Attalus,
who was collecting books for his royal library at Pergamos.
Here they suffered much from damp and cockroaches until
about the year 100 B.C., when they were sold at a great
price to the wealthy Peripatetic philosopher and book-

collector, Apellicon of Teios, who carried them back to
Athens. Apellicon died just before the capture of Athens
by Sulla in 83 B.C., and his library was seized by the conqueror and carried to Rome. There the manuscripts were
arranged by Tyrannion, a captive Greek, but then a freed-
man, a friend of Cicero and occupied in teaching. Thus it
came about that the works of Aristotle were edited and
published, much as we have them to-day, by Andronicus of
Rhodes. It was, perhaps, partly owing to the long burial
of the Aristotelian manuscripts at Scepsis that many forged
editions, imitations, and popular summaries of them made
their appearance.

Theophrastus's *History of Plants* seems to have consisted
originally of ten books, nine of which have been preserved
intact. The first deals with general anatomy and histology,
enumerating the external organs of a tree as root, stem,
branch, bud, leaf, flower, and fruit. It alludes to the sap,
fibres, vessels, flesh, wood, bark, and pith; distinguishes
woody from herbaceous plants; land from aquatic plants,
and permanent organs from those that are deciduous.
From the second to the fifth book is occupied with trees
and shrubs, with some reference to their diseases and to
the selection of timber. Here is described not only the
caprification of the Fig but also the artificial fertilisation
of Palms. The absence of seed in Elms and Willows is discussed, as is the dispersal of seeds by rain and floods; the
bleeding of turpentine from the stems of needle-leaved
trees; and the sensitive leaves of the *Acacia* in Upper
Egypt. The sixth book deals with shrubs; thorny plants;
those used for garlands; and the Roses, especially those
then employed in Greece for the manufacture of attar, *i.e.*
Rosa centifolia L. The seventh book deals mostly with the
vegetables of the kitchen garden; the eighth with cereals;
and the ninth with gums and exudations, and the means of
collecting them.

The second work of Theophrastus, known as *The Causes
of Plants*, was in eight books, of which six remain. Begin-

ning with the seed, the first deals also with grafting and budding and the effects of heat and cold. The second contains a discussion of the effects of weather and soil; and, while the remainder are largely devoted to the arts of cultivation, the work, as we have it, concludes with the diseases, death, tastes, and perfumes of plants.

It has been unjustly objected to Theophrastus's work that it evinces but little first-hand knowledge. He had not travelled, and he draws his information indiscriminately from philosophers, drug-collectors, farmers, woodmen, and charcoal-burners, but also, no doubt, largely from his many years of observation in his own large garden. Not being a physician he is not, like Hippocrates, interested in plants mainly as drugs. He gives lists of names with little or no descriptions, so that his work has become largely a peg upon which commentators have expended much ingenious speculation. In all he mentions some four hundred and fifty plants, among the more interesting being the *Nelumbium*, previously mentioned by Herodotus; the Papyrus, which is well described; *Ferula glauca* L., the Narthex of the islands of the Ægean; *F. Tingitana* L.; and several species of *Laserpitium ;* the Truffle; the Morel; several Fuci; "Thyion" from Cyrene; the Oleander; the Citron, and Black Pepper (*Piper nigrum* L.); the Banana; the Banyan; the Tamarind, and the Coconut from India. Thyion is described as resembling Cypress and having a very dense wood. It is the Thyine-wood of the Apocalypse, the favourite "Citrus wood" used by the Romans for the most costly table-tops, and the "Alerce" of which the woodwork of the Alhambra and the roof of Cordova Cathedral are constructed. Formerly known as *Thuya articulata* Vahl, or *Callitris quadrivalvis* Ventenat, it is now *Tetraclinis articulata* Masters. Its burrs come into the English market from Algiers as "Thoo'ee."

The Banana (*Musa sapientum* L.) may have been known to the Assyrians. Theophrastus describes it as "having leaves two cubits long, like ostrich feathers."

There can be little doubt that his " Koukiophora, a palm differing from the Date in having a round fruit of immense size, filled with delicious milk," is *Cocus nucifera* L., though it has been supposed to be *Hyphæne coriacea*. His description of the closing of the leaflets of the Tamarind at night, derived, no doubt—as were the remainder of his Indian descriptions—from the reports brought back by the scientific staff of Alexander's expedition, is the earliest account of nyctitropism, or " sleep " in leaves.

An examination of the *Historia Plantarum* shows that Theophrastus recognised Fungi, Algæ, and Lichens as plants, characterising plants in general as without the power of voluntary action or a moral sense. He divides plants into Flowerless and Flowering, though not using those terms precisely in their present sense, and treats flowers as leafy, *i.e.* petaloid, " capillary," or apetalous, as in the Grape, Mulberry, and Ivy, or both, one inside the other, as in the Rose. The calyx he considered as leaves; the ovary as the young fruit; but the stamens, with which the style was confused, as forming part of the flower as well as the corolla. He also divides plants into trees, shrubs, half-shrubs, and herbs, and distinguishes between annual, biennial, and perennial duration.

As we have seen, he deals with organs in the order of their development, noting, though without any magnifying glass, that the radicle or primary root is the first part of the embryo to enlarge in germination, and dividing all organs into those permanent and those deciduous. He recognised that the climbing organs of the Ivy are roots, as are also the dependent aerial structures of the Banyan and the parasitic organs of the Mistletoe, while some fleshy jointed underground structures are stems. Stems he divided into trunks, culms, and stalks, according to texture and prominent nodes. He distinguished leaves as stalked or sessile, opposite or scattered, simple or compound. He noticed the difference in the order of opening flowers between those inflorescences we term centripetal and those

that are centrifugal. He even observed the difference
between what we now call hypogynous, perigynous, and
epigynous insertion of the corolla, and between gamopetalous and polypetalous corollas. Equally interesting is
the fact that he recognised that the Thistles have a head of
numerous small flowers. Generalising the many variations
in the structures enclosing seeds under the term *pericarp*,
first used by Aristotle, he termed it, together with the
enclosed seed, the fruit. His distinction between angio-
spermous and gymnospermous plants is, however, very
different from the modern knowledge. In germination, he
says :

> " Wheat and Barley make their first appearance with
> only one leaf, Peas and Beans with several,"

no doubt combining the cotyledons with the first leaves of
the plumule. He also recognises the generally unbranched
stems and parallel leaf-veins of the plants we term Monocotyledons, as distinguished from the branched stems and
netted veins of Dicotyledons. Describing the interior of
the stem, he distinguished the bark (phloem), wood (xylem),
and pith, or rather protomeristem, and he compares this
with the epidermis, fibrous veins, and fleshy mesophyll or
inner tissue of leaves.

Among the natural groups that he recognised were the
culm-bearing, fibrous-rooted Grasses; Bulbous plants;
those with hollow stems and Umbelliferous flower-heads;
those with milky latex and crowded strap-shaped florets
(*Cichoriaceæ*) ; those with crowded flower-heads and spinous
foliage, such as the Thistles; and the cone-bearing and the
catkin-bearing trees. More than a hundred genera still
bear the names he gave them, though by our present rules
these are all quoted as of Linnæus.

Theophrastus seems to have kept careful records of those
periodic phenomena, such as leafing and flowering, that we
have in modern times dubbed phenology; and he notes
the effect of atmospheric drought and moisture in hastening

and retarding the fall of the leaf. In dealing with wild plants, as distinguished from cultivated plants, to which he pays more attention, he uses what are practically ecological groups, distinguishing, for instance, marine herbs (*Algæ*); submerged marine trees (mostly corals); marine shore-plants; those inhabiting deep fresh water and those of shallow lake shores, river-banks, and marshes respectively.

Modern writers, considering merely the comparatively small number of plants that he described, have disregarded these many important contributions to pure botany, which so fully justify Theophrastus's title of " Father of Botany." They were not the type of matter that would be quoted by Pliny; they were probably unknown to Albertus Magnus, but they bore fruit in the work of Valerius Cordus, if not also in those of Cæsalpinus and other botanists of the Renaissance.

Several of the works of Theophrastus must have been known to Cicero, who speaks of them as supplementing those of Aristotle; and he was one of the chief authors to whom Pliny was indebted for information about plants. Later on, however, he was but little known until Pope Nicholas V directed the Greek refugee Theodore Gaza to translate his works into Latin, which he did most admirably, publishing the result at Treviso in 1483. Manuscripts of Theophrastus exist in the libraries of Florence, Venice, Paris, the Vatican, Leyden, and Oxford, that at Venice being the most perfect. Aldus printed the Greek text with that of Aristotle's works on animals in 1497: Scaliger published *Commentaries* upon them at Geneva (in 1566) and *Criticisms* at Leyden (in 1584), and these were incorporated, together with Gaza's Latin version, in the excellent illustrated edition issued from Amsterdam by John Bodaeus a Stapel (in 1644). Of several subsequent editions that of John Stackhouse (Oxford, 1813–14) is, perhaps, the least satisfactory; that of F. Wimmer (Breslau, 1842) one of the most recent and best.

CHAPTER IX

GREEK SCHOLARS AT ALEXANDRIA

As we have seen, the interest of Hippocrates in plants was entirely medical: that of Theophrastus was more purely scientific. The latter's attitude towards the subject was philosophical, and he alludes contemptuously to the superstitious notions disseminated by the drug-collecting quacks of his time. These notions may have been deliberately used, however, as were the Phœnician travellers' tales, to enhance the value of their wares. One plant must be gathered at night for fear a woodpecker might pick out the eyes of the collector. When gathering another, it was necessary to draw a line three times around it with a sharp sword. Libations of wine must be poured on the ground before uprooting a third. If an eagle comes near whilst a fourth is being dug up, the herbalist will die within the year!

The marvellous tales associated with the Mandrake may have originated with these Greek quacks, although they have been stated to be of Chinese origin. The plant (*Mandragora officinarum Linné*), one of the *Solanaceæ* or Nightshade family and indigenous in the Mediterranean area, was early found to possess those narcotic qualities that placed it, in Shakespeare's language, among " the drowsy syrups of the world." Its repute as a love-philtre was, no doubt, increased, if not originated, by the fancied resemblance of its root to the body of a man or a woman. Pythagoras is said to have called it " Anthropomorphon "; Columella, the Latin writer on agriculture in the first century, styles it " semi-homo "; and Josephus tells many of the legends connected with it. It was said to be never, or very seldom, found wild except under a gallows, " where the matter fallen from a man hath given it the shape of a man," and that from the body of a woman had made the female plant.   " Its virtue is so mickle that it will immediately flee from an unclean man," so that it was necessary to "inscribe it with iron " directly it was seen, but not to touch it with the iron in uprooting it, but to use a staff of ivory.   It would shriek when uprooted, and human beings hearing it would, as Juliet says, " run mad " or shortly die.   It was, therefore, alleged to be the practice to tie an unfortunate hungry dog to the plant, and " cast meat before him, so that he may not reach it, except he jerk up the wort with him," the human collector meanwhile stopping his ears.   These ridiculous tales, " whether," as Gerard says, " of old wives or some runagate surgeons or physicke-mongers," led to the manufacture of elaborate counterfeits from the roots of the White Bryony (*Bryonia dioica* L.).   Matthiolus, in the sixteenth century, speaks of Italian ladies in his own time paying twenty-five or thirty ducats for one of these impostures.   Turner in his *Herball* (1568) writes:

> " The rootes which are conterfited and made like little puppettes and mammettes, which come to be sold in England in boxes, with hair, and such forme as a man hath, are nothyng elles but folishe feined trifles, and not naturall.   For they are so trymmed of crafty theves to mocke the poore people with all, and to rob them both of theyr wit and theyr money.   I have in my tyme at diverse tymes taken up the rootes of Mandrag out of the grounde, but I never saw any such thyng upon or in them, as are in and upon the pedlers rootes."

When Greece lost her liberty, she lost also the originality of her literature.   From the time of Alexander to that of Augustus, the great city of Alexandria, the site of which had been chosen with wonderful foresight by the great conqueror, became the centre of Greek intellectual activity.

The efforts of Attalus to make his library at Pergamos (for which the new writing material that derives from it the name of " parchment " was invented) a rival to that of Alexandria were unsuccessful. In the wealth of its endowments the " Museum " or college and library, founded in the Egyptian capital by the first Ptolemy, offered an attraction to learned men such as no other city could rival. Spontaneity had vanished from Greek literature, however, and laborious erudition became the keynote of the Hellenistic age. Although the great names of Euclid and Eratosthenes remind us that the achievements of this era were great in the mathematical and physical sciences, little that was new was contributed to biology.

In the sixteenth- and seventeenth-century herbals of Gerard and Parkinson we sometimes come upon quotations from Nicander's *Book of Treacles*, a name that carries with it a story belonging to this age. Nicander was hereditary priest of Apollo at Clarus near Colophon, living between 185 and 135 B.C. Of many works by him in prose and verse, two only are preserved. These are *Theriaca*, consisting of 958 Greek hexameters on the nature of venomous animals, the wounds inflicted by them and their cure, and *Alexipharmaca*, a similar " epic " of 630 lines on poisons and their antidotes. There is not much poetry in these productions and only little of botanical value, though 125 plants (thirty apparently for the first time) are mentioned either as antidotes or as poisons. The works gained, perhaps, in importance by being printed with the first edition of Dioscorides, the Aldine edition produced at Venice in 1499. *Therion*, being the Greek for " a wild animal," was used for a viper—as, for instance, in the Acts of the Apostles—and by an anticipation of homœopathy popular down to our own time, a confection of viper's flesh was held to be the most potent antidote against the viper's bite. It became known as " Theriake," which by Chaucer's time had become " *Triacle* " and a little later " *Treacle*."

From meaning this particular antidote the word came to stand for any antidote, then for any medicinal or any sweet syrup, to be restricted in our own time to the liquid molasses of sugar.

Poisons and their antidotes were popular subjects of study at this period, particularly, it would seem, among princes, so that a modern historian of botany heads one of his chapters, " The crowned poison-mixers." Plutarch is the authority for attributing an interest in poisons to Attalus III of Pergamos; but a better-known story is that of Mithradates the Great, the last King of Pontus.

> " He it was," says Pliny, " who first thought of drinking poison every day, with proper precautions, so that by becoming habituated to it he might neutralise its effects. He too was the discoverer of various anti-dotes, (one of which) is composed of fifty-four in-gredients, none used in exactly the same proportion. Pompey after defeating him found in his private cabinet a recipe in his own handwriting which taken fasting would render the taker proof against all poisons for that day."

This remedy was but a harmless mixture of Walnuts, Figs, leaves of Rue, and salt. According to the legend, when Mithradates wanted to take his own life, poisons had no effect upon him.

Important as is Strabo's work in the history of geography, there is nothing to show that he had any knowledge of plants. He represents the beginning of the period when Rome was supplanting Alexandria as an intellectual centre, and his Geography belongs to the later part of his life, that is to say, to some date subsequent to the birth of Christ. Beyond this account of the Papyrus and Nelumbium in Egypt, and of the Date groves and Balsam gardens of the plain of Jericho, it contains but little bearing on our present subject.

We need add nothing to what we have already said as

to Nicholas of Damascus—the supposed author of the treatise on plants ascribed to Aristotle—but can pass on to the mainly Latin writers of the later Roman Republic and early Imperial times, though we shall still have several to mention who employed the Greek language, such as Arrian, Galen, and, above all, Dioscorides.

CHAPTER X

We do not expect to find much original or pure science in the literature of Rome. The Romans were eminently practical people, and their earliest writers to make any mention of plants are dealing primarily with agriculture. Cato Major, the Censor (234–149 B.C.) in his *De Re Rustica* mentions 125 plants, including a good many cultivated varieties, as, for example, eight kinds of Vine. Varro, " the most learned of Romans " (116–28 B.C.), in his work with the same title, mentions only 107, including thirteen varieties of Olive. Virgil (70–19 B.C.), the greatest of Latin poets and, perhaps, the most poetical of all didactic writers in any language, mentions 164. Many species are, however, mentioned by more than one of these three, and the total number alluded to by all of them is only 245. Columella in his *De Re Rustica*—written, probably, during the reign of Nero, some forty years after the death of Virgil —mentions 260 species, including fourteen or fifteen kinds of Cabbage (*Brassica*), ten kinds of Figs, and fifty-six of Vine.

No one previous writer mentions half the number of plants enumerated by Pliny (Plate I*b*). His one surviving work, generally known as his *Natural History*, is almost entirely a compilation, and has been called a " repository of all the errors of antiquity "; but it is interesting to note that with regard to plants he makes some claim to original research. Premising that other writers have sometimes given paintings but generally only verbal descriptions, or even bare lists of names, of plants, he says that botanical knowledge is not difficult to obtain, and that he had himself been able to examine nearly all of the plants of which he writes in the garden of Antonius Castor. Of Antonius Castor and his garden little is known beyond what Pliny tells us in this passage. Castor, it appears, was a physician, who had died at over a hundred years of age before A.D. 77, and not only "cultivated vast numbers of plants with the greatest care," but also "enjoyed the highest reputation for his knowledge" of them.

Caius Plinius Secundus himself was born at Como, or possibly at Verona, A.D. 23. He visited Africa, Egypt, and Greece; served in the army in Germany, when he reached the shores of the North Sea in Belgium. He was appointed, by Nero, procurator in Spain; and, by Vespasian, præfect of the fleet then stationed at Misenum. He was thus near at hand at the outbreak of Vesuvius in August, A.D. 79. Going at once to investigate the phenomena of the eruption, he was overcome by the fumes of the volcano and died of suffocation at Stabiæ (Castellamare) whilst the neighbouring towns of Herculaneum and Pompeii were being destroyed. He was a man of prodigious industry, and during meals, or at any leisure time, he would have some book read to him, whilst he himself made notes or extracts. (It was one of his sayings that no book is so bad but that some good may be got out of it.) On leaving his bath he also had a book read to him whilst he was being rubbed down, or he would occupy the time by dictating to a secretary. Even for short distances in Rome he preferred to be carried in a litter to save time, and was always accompanied by his amanuensis with writing-tablets. In winter he carried gloves so that the cold might not interfere with his note-taking. Besides voluminous historical works, now lost, and his *Natural History*, he left behind him 160 closely written note-books.

The *Natural History*, also styled a *History of the Universe*, is an encyclopædic compilation in thirty-seven books, citing between 400 and 500 authors, Greek and Latin. It was compiled, as Pliny himself says, from some two thousand volumes, most of which are now lost. It was dedicated to his friend Titus, afterwards emperor, and was published apparently about A.D. 77—two years before the author's death. Although this immense work has a certain plan, it has no scientific method and evinces a total absence of the critical faculty. It has been said that it probably does not contain a single original observation in biology. As might be expected, there are not a few blunders and confusions, such as in the identification of the Holly and the Holm Oak.

Pliny's style is by no means always clear, and he seems often to have misunderstood his Greek authorities, or to have misrepresented them by his abbreviated quotations. It is recorded of him that he not only had the valuable faculty of falling asleep at any moment, but that he sometimes accidentally fell asleep in the middle of his studies, and unkind critics have suggested that this may in part explain some of his mistakes! His great storehouse of ill-digested knowledge has been called the earliest " popular " natural history book. It was probably the first to be printed, being issued in 1469 from Venice by John of Spires. Cuvier speaks of the work as " one of the most precious monuments that have come down to us from ancient times, affording proof of an astonishing amount of erudition." It cannot be denied that, in spite of its defects, Pliny has preserved for us a vast mass of information that is no longer available elsewhere.

Plants occupy a large portion of the work—16 of the 37 books. Book xii is on trees; xiii on their products, fruit, gums, perfumes, etc.; xiv on the Grape and the making of wine; xv on the Olive, Fig, Apple, and other fruits; xvi on forest trees and timbers; xvii on their culture, diseases, pruning, training, etc.; xviii on farming and cereals; xix on horticulture; xx on the medicinal properties of garden plants; xxi on flowers, bees, honey, and plant-anatomy; xxii on herbs used in medicine and in cookery; xxiii on the medicinal properties of cultivated, and xxiv on those of forest trees; and xxv, xxvi, and xxvii on the medicinal properties of wild herbaceous plants. Mention is made of nearly a thousand plants—more than were to be dealt with again by any writer until the Renaissance. In 1583 Cæsalpinus writes only of about 1500, though Caspar Bauhin forty years later enumerates 6000, Tournefort by 1700, 10,000, and Ray by four years later, 18,000; whilst at the present day more than ten times that number are known to science. Alexander von Humboldt, whose *Cosmos* (1844) has often been compared to Pliny's work, writes of the latter:

> "The want of success that has attended Pliny's undertaking is to be ascribed to his incapacity of mastering the materials accumulated, of bringing the descriptions of nature under the control of higher and more general views, or of keeping in sight the point of view presented by a comparative study of nature."

As an example of his method we may quote Pliny's treatment of the question of sex in plants. He seems to have made some advance, in that he recognises the existence of sex in all plants, herbs as well as trees; but the only example he gives is the old one of the dioecious Date-palm.

> "In a forest of natural growth," he writes, "the female trees will remain barren if isolated from the male, and several females may be seen bending towards the latter, with foliage of a softer character. The male tree, on the contrary, bristling with erect leaves, fecundates the others, by its presence, by its exhalations, and even by the dust it emits. When, too, it is cut down, the females, reduced to a state of widowhood, become barren. So clearly, indeed, is this sexual union recognised as taking place, that the notion has arisen of securing the act of impregnation by man's agency, the blossoms from the male trees being gathered for that purpose, and even sometimes nothing more being done than to sprinkle the dust taken from the same over the female trees."

Apparently about ten years after Pliny's death (*i.e.* about A.D. 90) there was composed an interesting little anonymous Greek work known as the *Periplus Maris Erythræi*, or Route-book of the Red Sea. It appears to be the work of an Alexandrine Greek merchant, writing with a purely practical aim and mentioning the chief exports and imports of each port.

After passing Aualites, the modern Zeila, or Auzal of the Somalis, mention is made of an unlocalised Mosyllum, the chief port for the export of Cassia, and other ports towards Cape Aromata (now Guardafui), whence Frankincense, Myrrh and tortoise-shell were bartered for glass, gold and silver plate, or Roman coins. From them there seems to have been a direct trade with India, and although nothing is said of any importation of Cinnamon, it is expressly stated that the large size of Alexandrine vessels trading with the ports was due to their large export of Cassia. Ninety miles beyond Guardafui was the port of Opone, which imported from Barygaza (Baroch, at the mouth of the Nerbudda), Corn, Rice, and the "honey" produced by a reed called "sacchari." Pliny mentions sugar as a drug, but this is the first reference to it as an article of commerce. Some authorities, however, insist that this sugar was the *tabachir*, or opal, found in the Bamboo.

On the opposite coast, the south-east of Arabia, 225 miles east of Arabia Eudæmon (Aden), was Cane (Hisn Ghorab), where the Frankincense of Hadramaut, the ancient Saba or Sheba, was the chief export. Beyond was the lofty mountainous headland of Syagrus (Cape Fartak), which had then become, since the first adventure of Hippalus, the starting-point for direct sailing before the south-west monsoon for India. From the large and fertile island of Dioscorides (Socotra) was obtained the Indian "cinnabar," the dark red resin of *Pterocarpus Draco*, now known as Dragon's blood. At a depôt called Omana, apparently on the coast of Baluchistan, Sandalwood—of which this is the earliest certain mention—and Ebony were exchanged for the pearls of the Persian Gulf. Among the articles of commerce at the seven mouths of the Sinthus (Indus) was Indigo, and from Barygaza, Nard and other perfumes. Pepper was then, as now, a leading export from ports further down the Malabar coast and also from Palæsimundus (Ceylon). There is no mention of Cinnamon among the exports from that island, however, and, as there is no mention of it as the produce of Ceylon in Arabic or Oriental writers until the thirteenth century, Sir Emerson Tennent suggested that the tree was not indigenous. This suggestion has not, however, received the assent of botanists.

CHAPTER XI

AMONG the greatest treasures of the Imperial Library at Vienna is a wonderfully perfect early manuscript of a work that had immense influence upon botanical study, although but little is known as to its author. The name of Dioscorides is not among the many authorities cited by Pliny, nor is Pliny's great work quoted by Dioscorides. This by itself suggests that they were practically contemporaries.

Pedacius Dioscorides, a native of Anazarba in Cilicia, was born probably about A.D. 40. After serving as a soldier he devoted himself to the study of physic. Although he travelled in Asia Minor, Greece, Italy, and Provence, collecting many plants in those countries, the bulk of his descriptions are made on the authority of others. He quotes Theophrastus, Nicander, and other—mostly Greek—writers, although not nearly as many as does Pliny. He gives Latin, Dacian, Gallic, Punic, and Egyptian synonyms for some of his plants. His work, which is professedly a *Materia Medica*, deals only with some 600 species, and his descriptions are often extremely meagre, or even absent altogether. There is some slight attempt at an orderly sequence of plants, the main grouping being into aromatic, alimentary, and medicinal species. Occasionally we find a series of *Labiatæ*, *Umbelliferæ*, *Compositæ*, *Boraginaceæ*, or *Leguminosæ*, showing that he was not blind to natural affinities. In spite of their superficial similarity in foliage, he recognises the fundamental differences between the Stinging Nettle and the Dead Nettle.

The work was probably written about A.D. 70 or 80, and

77

the numerous perfect manuscripts of it that exist show that it was valued before the invention of printing. There are three manuscripts at Leyden; a fine one of the ninth century at Paris—probably written in Egypt, since, as well as figures, it has some Arabic and Coptic names inserted. Two at Vienna are of the greatest importance, however, and that known as the Neapolitan is considered the older. The other Anician Codex, as it is termed, is the more interesting, for it bears the name of the princess Juliana Anicia, daughter of the Emperor Flavius Anicius Olybrius. Anicia, who was an ardent Christian and builder of churches, survived the accession of her cousin Justinian to the Imperial throne in 527. The manuscript is on vellum and is written in uncial Greek characters showing the characteristics of early sixth-century script. It is illustrated by a number of brush drawings, many of which are extremely faithful; and coloured probably when the manuscript was first written. Among them appears the Mandrake with the dog duly tied to its root.

In the sixteenth century Augier Ghislen de Busbecq (1522–92), a Fleming and the friend and correspondent of Clusius and Mattioli, was twice sent by the Emperor on embassies to Soliman II at Constantinople. He was a most painstaking collector, and it is said that to him our gardens are indebted for the Tulip, the Lilac, the Mock Orange (*Philadelphus coronarius* L.), and the Horse-chestnut. He brought back to Vienna a fine collection of Greek manuscripts, among them the above-mentioned Anician Codex of the *Materia Medica* of Dioscorides. Busbecq found it in the possession of a Jew, who demanded for it a hundred ducats—more than the ambassador had in his purse. His friend Mattioli was then engaged upon the Commentaries upon the work of Dioscorides, which he published in 1548, and Busbecq accordingly persuaded the Emperor " to redeem such an illustrious author from that servitude."

It was his inspection of this beautiful manuscript in 1784 that decided Dr. John Sibthorp to undertake his *Flora Græca* as illustrative of the work of Dioscorides. Plates were drawn from the manuscript under the care of the elder Jacquin (1727–1817). From these, however, only two copies seem to have been printed. One was sent to Linnæus and thus came into the possession of the Linnean Society of London, where it remains. The other was sent, either on loan or as a gift, to Sibthorp, who died in 1796. This copy is now in the Botanical Library at Oxford. In 1906 a photographic monochrome facsimile of the Codex was published, and a copy may be seen at the Natural History Museum at South Kensington.

At the Renaissance the works of Dioscorides were considered of the highest authority and no drug was genuine that did not agree with his description. Unfortunately, botanists in Northern Europe, ignorant of the geographical limitations of species, tried to identify the plants of their own lands with those that he had described from the Mediterranean area. The *Materia Medica* was translated into Latin by Hermolaus Barbarus (1454–93)—a Venetian ecclesiastic of high rank, who afterwards published criticisms of Pliny —and printed at Medemblik in Holland in 1478. The original Greek was first printed by Aldus at Venice in 1499, together—as we have seen—with Nicander's poems; the Greek and Latin together, by Vergilius at Cologne in 1529.

Pietro Andrea Mattioli, who was born at Siena in 1501, and died at Trieste in 1577, was a physician who, in 1544, published an illustrated volume of *Commentaries on Dioscorides* in Italian. Four years later he published a translation into Latin by himself. This work was translated into various languages and ran through many editions, 32,000 copies being said to have been sold before 1561. Obviously so popular a work must have exerted an immense influence on the progress of botanical study. The effect of this and other similar and nearly contemporary works was unhealthy, however, in that it encouraged the waste of an infinity of labour upon the attempt to identify the vaguely described plants of Dioscorides, thus diverting the minds of botanists from the study of plants. As Pulteney wrote in 1790:

" It is a mortifying reflexion in the annals of human knowledge, that the bulk of these learned men, after their immense labours, mistook, in numberless instances, the road to truth, and did but perplex the science they wished to enlighten. The descriptions of plants in the antient authors, were, at best, short, vague, and insufficient; and with this inconvenience, the study of nature herself was neglected."

On the whole, the best editions of Dioscorides are those of J. A. Saracen, with a Latin translation, Frankfort, 1598, folio, and of C. Sprengel, in two octavo volumes, 1829–30.

Dioscorides seems to have had a better notion of specific distinctions than any of his predecessors. In spite of his meagre descriptions, we realise that he recognises some of that great variety of Labiates, of Poppies, and of species of *Euphorbia* and *Artemisia* that is characteristic of the Mediterranean flora. He mentions the Aloe (*Aloe vulgaris* Lam.) as growing in the islands of Hyeres (Staechades), and the Chervil (*Anthriscus Cerefolium* Hoffm.), curiously omitted by Theophrastus, though long cultivated at Athens. Aristophanes in his *Acharnians* taunts Euripides as having been a greengrocer who sold Chervil for his mother. It is enlightening as to the vagueness of the early ideas of sex in plants that Dioscorides thinks the Scarlet Pimpernel to be male and the Blue species female. Among the most interesting of all the plants mentioned by him are Sugar and Rhubarb.

We learn from Strabo that Nearchus, who commanded Alexander's fleet on its return from India, had spoken of the Indians as making honey from a reed without the aid of bees, and the poet Lucan, who died before Pliny, writes of them as drinking the sweet juices from the soft reed. Pliny had access to the account of this voyage by Onesicritus, or to an abstract of it by Juba, king of Mauritania, and he writes of it as:

> " Saccharon . . . a kind of honey, which collects in reeds, white, like gum, and brittle to the teeth, the larger pieces . . . about the size of a filbert . . . used only in medicine."

Dioscorides compares it to salt, while the author of the *Periplus of the Erythræan Sea*, as we have seen, speaks of it as an export from Opone to Arabian ports.

The art of boiling sugar was introduced into China from Bengal early in the seventh century, but it was not until the thirteenth that they received from Egypt the art of refining the sugar by the use of ashes. About A.D. 650, as we learn from the *Geography of Armenia*—falsely ascribed to Moses of Chorene—sugar was grown and manufactured at Gundé-Sháhpúr, near the Euphrates, in Khuzistan. Under Arab dominion it spread from India to Sus in Morocco, to Sicily and Andalusia. In the Middle Ages the best sugar was made in Egypt, and to this day coarse sugar is known in India as Chinese, refined sugar as Cairene. Venice was the depôt for the supply of Europe with Egyptian sugar, the name " sugar " having come into our European languages from the Persian *shakar*, through the Arabic. Spain took up the work of dissemination from the Arabs, introducing the Sugar-cane to Madeira in 1420, and to the West Indies before the close of the fifteenth century. Sugar remained a luxury after the time when Queen Elizabeth blackened her teeth by her fondness for it, and as late as the days of Gilbert White's career at Oxford a sugar-loaf was a polite and valuable present. The amount of sugar used in Great Britain in 1700 was 10,000 tons; in 1800, 150,000, in 1900, 1,624,000 tons. The greater part of this last huge amount was the product, not of the Sugar-cane, but of the Beet.

Dioscorides describes the drug *Rha* as a blackish root, reddish within, from the regions beyond the Bosphorus. Ammianus Marcellinus in the fourth century explained that it grew on the banks of the river of that name, the Volga, where *Rheum Rhaponticum* L. actually occurs. The drug had long been known in China, and by the eleventh century Chinese Rhubarb was known to be superior to the *Rha barbarum*. Marco Polo writes:

> "All over the mountains of the province of Tangut Rhubarb is found in great abundance, and thither merchants come to buy it, and carry it thence all over the world."

Here too in the extreme north-west of China *Rheum palmatum* L., the present source of the drug, was found wild by Colonel Prejevalsky in 1872. In the fourteenth century it reached Europe, partly by way of India and Alexandria, under the name of "East Indian Rhubarb," and partly overland to Aleppo and Smyrna under the name of "Turkey Rhubarb." The difficulty of preserving the drug from insect attack during its long overland transit made it a costly article. In the sixteenth century it is quoted at ten times the price of Cinnamon and four times that of Saffron, and a century later at sixteen shillings a pound. From 1704 to 1860 it was a monopoly of the Russian Government, coming from Urga or Kiakhta to Moscow, and known as "Russian Rhubarb"; but it now comes by sea to England and bears the true name of "Chinese Rhubarb."

The only Greek writer subsequent to Dioscorides who must be mentioned here is Galen. This great physician was born at Pergamos A.D. 130 and became the medical attendant of Marcus Aurelius and Lucius Verus, living till about the year A.D. 200. Those of his very numerous works that survived exercised an enormous influence upon the study of human anatomy and physiology, dietetics and hygiene. He extended the Hippocratean theory of humours by that of temperatures, which he applied to drugs as well as to the human body. Most of what he wrote as to plants is either borrowed from Dioscorides or only adds a few mistakes to previous errors of identification.

CHAPTER XII

THAT very curious book, the *Christian Topography* of Cosmas—always known as Indicopleustes, the navigator of India—contains a few items of interest to us. It was written in Greek at Alexandria about A.D. 540 by a monk, probably a Nestorian, who had been a merchant. It was preserved in two manuscripts, one of which, dating from the eighth century, is in the Vatican Library. The author mentions, obviously from personal knowledge, numerous small islands (probably the Maldives) near Taprobane (Ceylon) that produce Coco-nuts; the existence of Nestorian churches and Pepper on the Malabar Coast; Cloves grown somewhere between Ceylon and China (Tzinista), and the incense and spices of Somaliland.

In A.D. 552 two Persian monks explained to the Emperor Justinian that the silk of China was not combed from trees but was spun by a caterpillar. They made a journey to that distant land and returned with some silkworms' eggs and Mulberry-leaves concealed in a hollow tube. In the same century Egyptian Papyrus was brought to Marseilles.

Driven from Constantinople in A.D. 431, the Unitarian followers of Nestorius, an enthusiastic student of Aristotle and other Greek philosophy, spread themselves over Persian territory in south-eastern Asia, established many monasteries and schools and translated many of the Greek and Latin philosophical classics into Persian and Syriac. In its study of mineral and vegetable drugs their medical school at Edessa in Mesopotamia was a prototype of the later Italian schools of Monte Cassino and Salerno, but was dissolved by the orthodox Zeno in A.D. 489. In conjunction with

Jewish scholars they founded a medical school at Gundé-Sháhpúr, and it was at the Nestorian monastery of Bozrah on the confines of the Syrian desert, that Mohammed, the young nephew of the head of a caravan from Mecca, received his education.

When we come to that remarkable historical phenomenon —in which a hitherto obscure race of nomads, under the stimulus of a new religion, burst forth upon the ancient and decadent civilisations from the Pillars of Hercules to the Indus and imposed their own beautiful and flexible language upon other races and even other religions—we do not find at first that enlightened appreciation of science that afterwards distinguished many of the rulers in Islam. It was not until the division of the caliphate in the eighth century between Bagdad and Cordova that the obscurantism that had characterised early Mohammedanism—as it had occurred before with early Christianity and was to occur long after with early Protestantism—was to give way to a love of knowledge for its own sake.

In the Medicean Library at Florence there was an Arabic manuscript compendium of medicine, with a treatise on the properties and names of plants used as remedies, compiled as early as A.D. 718 from Hippocrates by Ahmed Ben Ibraham, physician to the Ommayad Caliph Yazid II. In the Royal Library in Paris there was another, dated 743, by a monk named Ebn Abu Zaher, who seems to have travelled extensively. The Caliph Abdurrahman I (A.D. 755–88) laid out a botanical garden at Cordova, caused rare seeds to be collected by travellers in Syria and other parts of Asia, and planted the first Date-palm to be grown in Spain.

It was not until the Abbaside Caliphs had transferred their capital from Damascus to Bagdad that the golden age of Arab learning really began. The early translators of Greek works into Arabic were more often Arabic-speaking Christians, chiefly Nestorians, rather than Moslems. The Nestorian scholar found his way into the household of the wealthy Moslem as a tutor, while the Jew was similarly tolerated as a physician. Haroun-al-Raschid, the friendly correspondent of Charlemagne, placed his schools under the direction of a Nestorian and introduced the Homeric poems to his people. His son and successor, Al-Mamun (813-32), collected from Constantinople, Armenia, Syria, and Egypt all the Greek manuscripts he could purchase, and employed a number of scholars to translate them into Arabic. This took place just about the time when our own Alfred was translating Latin works into the Early English tongue of his subjects.

Abu Zakaria Jahia Ben Masawaih, otherwise known as John the son of Mesue, or Mesue the elder—a Syrian Christian, whose father was an apothecary at Gundé-Sháhpúr—became physician to Haroun and his successors. He was specially commissioned to edit old medical works, including those of Galen and Aristotle. His amanuensis, Abu Zeid Honein, known as Honein or Johannitius—also a Christian and the son of an apothecary—visited Greek towns to study the language, and Bassora to perfect his Arabic. He then practised at Bagdad, becoming physician to the Caliph Al-Motewekkil. Although Mesue the elder died in 857, and Honein in 873, the latter's family continued the work of translation. By them Hippocrates, Galen, Dioscorides, and the treatise on plants ascribed to Aristotle were rendered into Arabic, as were many other Greek scientific works.[1]

From first to last the Arabic school made no claim to originality, entertaining an almost superstitious veneration for their Greek teachers. With them the deification of Aristotle had its beginning, and from Aristotle they derived a fatal love for attempting the whole circle of the sciences.

One of the first of the Moslem scholars was Abu Bakr Mohammad Ben Zacharia Arrazi, better known as Al-Razi —so named from Rai, in Khorasan, his birthplace. He has been called the Galen of his age, and studied in Bagdad,

---

[1] Many of the early manuscripts of Dioscorides have marginal notes in Syriac.

became director of its hospital, and probably died there
A.D. 932. Among the two hundred and more works by
him, of which the titles come down to us, is one on aromatic
plants and another on edible fruits. In 1531, Otto Brunsfels
published at Strasburg a volume containing " the very
useful little work on simples by Razi, son of Zacharias,"
together with works on the same subject by John Serapion
and Averroes, and one ascribed to Galen on Centaury.
This Serapion (whose work, known in its Latin translation
as *De Medicamentis Simplicibus*, is in fact a herbal) was the
second of that name and belongs apparently to the end of
the eleventh century, Serapion the elder being a Syrian
physician who was a contemporary of Al Razi. A bar-
barous Latin translation, made at the end of the thirteenth
century by Simon Januensis and Abraham of Tortosa, was
practically the pharmacopœia of the Middle Ages. Printed
versions appeared at Milan in 1473 and at Venice in 1552.
It deals in 365 chapters with vegetable remedies and—as is
characteristic of Arabic pharmacy—treats also at length of
those of mineral origin.

Mesue the younger was a Jacobite Christian, born on the
banks of the Euphrates. He studied at Bagdad and went
afterwards to Cairo, to the court of Hakim II, accompanying
him to Andalusia, where he died about A.D. 1015. Hakim
is said to have had a library of 400,000 volumes, and to have
welcomed scholars of all nations and creeds. Mesue left a
treatise on simples, and this, translated into Latin, was
printed in no less than twenty-six editions in the fifteenth
and sixteenth centuries. It was employed by Sir Theodore
Mayerne in the preparation of the first London Pharmacopœia in 1618.

The greatest names undoubtedly in the history of Arabic
philosophy are those of Ebn Sina, universally known as
Avicenna, and Ebn Roschd, known as Averroes. They are,
however, less important in the history of botany than in that
of medicine or metaphysics.

Avicenna was born in Bokhara about A.D. 980 and seems

G

to have been a youthful prodigy in his powers of assimilation.
For instance, before he was seventeen he had read the
Metaphysics of Aristotle forty times, so that he knew them by
heart. He was indefatigable in his studies by day and night,
and when baffled by difficulties betook himself to the
mosque to seek enlightenment in prayer. Lecturing on
Logic and Astronomy and acting as physician to various
emirs, he was also a voluminous writer. When he died at
Hamadân, in 1037, he left upwards of a hundred works.
The logical method of his *Canon* made it a university text-
book at Montpellier and Louvain down to the middle of
the seventeenth century, and earned for its author the title
of Prince of Physicians. Beyond general views of plant life
borrowed from Aristotle and an enumeration of medicinal
plants, it contains no botany. None of his other works deals
with that science, although he is said to have made use of
coloured drawings of plants in teaching.

Averroes, born at Cordova in 1126, became Kadi of
Seville in 1169, under the patronage of the Caliph Jusuf
Almansur. He died at Morocco in 1198, his life-work being
the exposition of Aristotle. On to the teaching of his Greek
master he grafted Neoplatonist idealism, and this led him
to conceive that the world was an emanation from the all-
pervading intellect of God. Rejected by Moslems as a
materialist or atheist, he found a hearing among the Jews of
Provence, with the anti-clerical Frederick Barbarossa and
in the universities of Northern Italy. The Dominican
philosophers Albertus Magnus and Thomas Aquinas vin-
dicated the orthodoxy of Aristotle against his glosses and,
with the revived study of Greek, at the close of the fifteenth
century, Averroism declined.

Abd-Allatif of Bagdad (1162–1231) seems to have possessed considerable knowledge of plants. Under the protection of Saladin he visited Egypt and wrote an account of
the country, which account has been translated into French.
He describes as peculiar to Egypt the Okra or Gumbo
(*Hibiscus esculentus* L.) ; the wild Sycamore Fig (*Ficus*

Sycomorus L.), the cutting of its fruit before ripening and the use of its timber; the Balm of Gilead (*Commiphora Opobalsamum* Engl.); and the Kholkas (*Colocasia antiquorum* Schott), with acrid rhizomes rendered edible by boiling.

The thirteenth century Ibn Baithar, or Ebn-al-Beitar, was of considerably more consequence as a botanist than any of the Arabic writers we have named. He was born at Malaga, and travelled in search of plants in Tunis, Egypt, Greece, Syria, and even as far as Medina and Mosul. After holding high medical and scientific office at the Egyptian court, he died at Damascus, A.D. 1248. It was claimed for him that he described no plants that he had not seen, and those that he describes (adding their Greek and Spanish names) amounted to about 1400, as against Pliny's 1000. His work remained in manuscript at the Escurial and in the libraries of Madrid, Paris, and Hamburg until 1833, and was somewhat unsatisfactorily edited by Sontheimer in 1840–2.

Impressed, no doubt, by their achievements in chemistry and physics, Humboldt and others have belauded the Arabs as the founders of modern science, because, not content with observation or even with measurement—both of which appear in Greek science—they had resort to experiment. In botany, however, as we have seen, with the one possible exception of Ibn Baithar, they neither experimented nor observed, but were content to copy the Greeks.

If the scientific writers in Arabic are disappointing from our point of view, their travellers, devoted mainly to commerce and with no pretensions to science, are more fruitful of novel information. By the close of the seventh century their conquests extended to Kashgar and the Punjab, and throughout Northern Africa. Within a century and a half they were trading simultaneously with Northern Europe, Madagascar, India, and China. No other race affords more striking examples of extensive land journeys undertaken by private traders.

Centuries before Mohammed, Arabs seem to have crossed the Red Sea and settled in Abyssinia. At a date scarcely less remote they are credited with the working of the gold of the Sofala coast and with trade connections with Persia, India, and even China. The incense of Yemen, the ancient Saba, had gained for that part of the peninsula the name of Arabia Felix and had formed the staple of a trade that, as we have seen, was Phœnician and Egyptian before it was Greek or Roman. On the south-east of the peninsula, Gerrha had been an important emporium of Indian commerce from Phœnician times, and from it radiated caravan routes across the desert to the prehistoric copper-mines of Sinai, to Tadmor and to Damascus.

Soleyman the Merchant, before A.D. 851, had reached China from Bassora, by way of Muscat, the Malabar Coast, the Maldives, Ceylon, the Andamans, and apparently the Straits of Malacca. In his narrative we have the following first account of Tea:

> " A certain herb, which they drink with hot water, and of which great quantities are sold in all the cities: it grows on a shrub more bushy than the Pomegranate, and of a more taking smell, but with a kind of bitterness. The Chinese boil water, which they pour upon this leaf, and this drink cures all sorts of diseases."

The editor of this narrative, Abu Zeid Hassan of Siraf, adds much interesting matter. The sea between India and China has gulfs from whose shores are obtained Ebony, Redwood, Brazil-wood, Aloes, Camphor, Nutmegs, Cloves, and Sandalwood. Brazil-wood was probably *Cæsalpinia Sappan* L.; and the Camphor, here mentioned for the first time, may have been that of *Dryobalanops aromatica* Gærtn., from Sumatra. The Coco-nuts, Sugar-cane, Bananas, and Palm-wine of the Nicobar Islands are also mentioned.

About A.D. 880 Ibn Khordadbah, Director of Posts and Police in Media, drew up for the Caliph the driest of official reports on trade routes. It contains, however, one most interesting passage on the course of trade from west to east, which was in the hands of Jewish and Russian merchants.

The Jews, we are told, spoke Greek, Latin, Persian, Arabic, the Frankish dialects, Spanish, and Slav. From the ports of France and Italy they sailed to the Isthmus of Suez and passed down the Red Sea to India and Farther Asia. Going to the Syrian coast and up the Orontes to Antioch, they descended the Euphrates to Bassora and the Persian Gulf to Oman. The Russians brought fox and beaver skins to the Mediterranean. Descending the Volga, they crossed the Caspian and brought their goods on camels to Bagdad.

Although not described by Ibn Khordadbah, there was from the eighth to the eleventh century a trade route from India through Kharism to Novgorod, Wisby in Gothland and Sweden, as is evidenced by the large number of coins from Khorasan, belonging to the tenth century, which have been found in Sweden.

Spices and perfumes were the staple of Arab trade. Incense was not only in demand for worship, but—with other powerful aromatics such as the Opopanax, mentioned by Dioscorides—was desirable to counteract the odours resulting from the absence of sanitation and personal cleanliness. Pepper and other condiments were even more necessary to season the salted food of winter and the salted fish of Lent, both often very imperfectly cured before the use of saltpetre became general.

As Arab medical science pervaded Europe, the Arab drugs from the West were in request. As communication with the East was increased, it was found that the looms of Europe could not equal the muslins of Mosul or the calicoes of India. Alexandria was the centre of all this trade, the Arabs having the monopoly of its eastern portion, and the Italians—and eventually the Venetians—almost exclusively commanding the European side. The spices of the East reached England, whether from Alexandria or Aleppo, mostly by way of Venice, Nuremberg, the Hanse towns, Antwerp and Bruges, until the great trade revolution of the close of the fifteenth century that resulted from the rounding of the Cape by the Portuguese and the discovery of America

In what has been happily called the Sindbad saga, belonging originally to the eighth or perhaps early part of the ninth century, we have a collection of all that was most marvellous in Arab travel attributed to a kind of Arab Ulysses. Though embellished with wonders from Homer, Pliny, and other sources, it nevertheless contained a considerable nucleus of truth. All Sindbad's seven voyages seem to have been in the Indian Ocean. In the first he reaches an island (which may well have been Sumatra) rich in Camphor and Pepper, the former being obtained by boring a hole in a tree and collecting the sap. The roc of the second voyage may possibly have been the quite recently extinct gigantic bird Æpyornis of Madagascar. The filthy hairy dwarfs of the third are, doubtless, ourang-outangs of some of the islands of the Eastern Archipelago. The island of Sandalwood may have been Timor, as the neighbouring Clove islands, with their sea-cows and sea-camels, were the Moluccas, where dugongs might then have occurred. The Pepper and Coco-nuts of the island on which he was wrecked on his fourth voyage might have been one of the Andamans, as Sindbad was then trading in Pepper, Lign-aloes and pearls, off what seems to be the Coromandel coast. The Old Man of the Sea, silent, fruit-eating, with immensely strong legs and skin as shaggy as a buffalo, is clearly one of the large apes, and his association with Vines suggests Banda in the Moluccas. The sixth voyage is in the main an accurate account of Ceylon, with its jewels, Camphor and Lign-aloes, such as might be compiled from Ibn Khordadbah and Pliny. The fragrant light-coloured resinous wood of species of *Aquilaria* is in Sanskrit *Agaru*, in Arabic *Aqulugin* or *Agalluchi*, and in Latin *Agallochum*. This name was corrupted into Aloes-wood or Lign-aloes, and in various European languages into *Eagle-wood* or its equivalents. It retains its fragrance for years, is burnt in Indian temples or used in inlaying, and sells in Sumatra at £30 per cwt.

Considering the wide range of their commerce and the industry of their medical writers, which extended over several centuries, the Arabs added comparatively little to our knowledge of plants. Such names as *Alchemilla* (the Lady's-mantle), *Alkekengi* (the Winter Cherry), and *Berberis* still bear witness to their botanical work. The leaves of the Senna (*Cassia* spp.), the fruit of the Tamarind (*Tamarindus indica* L.), and the seed and aril—or fleshy seed-coat—of *Myristica fragrans* Hout., Nutmeg and Mace, from the Banda Islands, were the chief vegetable drugs that they added to the pharmacopœia.

The Clove may have been known even earlier, perhaps by Chinese commerce with Arabia before Arabs had ventured far afield. There are few more curious stories of plant-names than that associated with this plant. Pliny writes:

> " There is in India a grain that bears a considerable resemblance to pepper, but is longer and more brittle: it is known as *Caryophyllon*, and it is said to be grown in a sacred grove in India: with us it is imported for its aromatic perfume."

Whether this was or was not the Clove, there is no doubt as to the meaning of the name *Caryophyllon*. *Karuon* is one of the Greek names of the Walnut, being supposed to refer to the brain-like convolutions of the seed within the nut. We now retain *Carya* as the name for the Hickories of America, near allies of the genus *Juglans*. The plant yielding the Indian spice was supposed, therefore, to have a leaf like that of the Walnut. Paulus Ægineta, a celebrated surgeon of Ægina, writing in Greek, apparently in the seventh century, seems to be the earliest writer to describe the plant we know as the Clove under the name *Karuophullon ;* but as early as the time of Charlemagne this had become corrupted to *Gariofilas*. Albertus Magnus writes *Gariofilus*, and later we have the Italian *Garofalo* and the French *Giroflée*.

While botanists long kept the name *Caryophyllus aromaticus* L. for the Oriental Myrtaceous tree (now known as

Eugenia caryophyllata Thubnerg and a native of the Eastern Archipelago, and whose flower-buds afford the once costly spice), these names were transferred in mediæval times to a cheap substitute, the leaves of which certainly do not resemble those of the Walnut. The dark brown inferior ovary surmounted by the little spreading calyx and dried inrolled petals of *Eugenia* naturally suggested comparison to a nail, the French *clou*, whence comes the present name " Clove." The Carnation, whose flowers have the same perfume as the spice, became known as Gilliflower, a corruption of *Giroflée*, and as the *Clove Carnation*. Chaucer writes of

> " many a clove gilofre
> And notemuge to put in ale ";

and this cheap substitute for Cloves from our flower-gardens had also the name of *sops-in-wine*. Thus the name of some spice-bearing plant with leaves resembling those of the Walnut, the nut of Jupiter (*Juglans*), became transferred to the flower of Jupiter (*Dianthus*), and a family, many of which have narrow grass-like foliage, now bears the name *Caryophyllaceæ*.

CHAPTER XIII

THE SCHOOL OF SALERNO

FROM the fifth to the ninth century, the period of barbarian inroads, the record of literature and science in Europe was dark indeed. Although the knowledge of Greek had practically disappeared from the Western Empire by A.D. 600, it is probable that the race of learned physicians never died out utterly, their text-book being Discorides in a Latin or an Arabic version.

To realise the low ebb to which botany sank, even with the majority of the scholars of the age, we have only to look at the *Herbarium* of Apuleius Platonicus. We know nothing of the author and but little of the early history of the book. Apuleius was certainly not the Platonist and rhetorician of Madaura in the third century and the author of the *Golden Ass*, but belonged probably to the fourth. His work is sometimes entitled *De Herbis, sive de Nominibus ac Virtutibus Herbarum,* or later *Herbarium Apuleii Platonici quod accepit ab Escolapio et Chirone Centauro magistro Achillis.* The book gives the names of about 180 medicinal plants in Greek, Latin, Egyptian, Punic, Celtic, and Dacian, some of which are named in Oriental languages. There is also a short description of each, the place of its growth, its properties, and the diseases to which it is applicable. The book is full of errors in identification, however, and the rules for the administration of the simples partake as much of magic as of medicine. The mediæval title of the book suggests that it came from the Eastern Empire. It owes much indeed to Dioscorides, and the numerous existing manuscripts of it suggest that it was the common text-book of those practitioners whom Fuchs afterwards described as the " *vulgus herbariorum.*"

Several of the manuscripts are illustrated with very crude coloured drawings, apparently copied from some ancient originals, and when the book was printed in the fifteenth century these were reproduced. One of their most noteworthy characteristics is that when a plant was supposed to heal the bite of some noxious animal, the animal is also figured. The imaginary likenesses between parts of the plant and of the animal are not brought out, however, as they were later in accordance with what is known as the doctrine of signatures.

Botany at this time had sunk to the level of the mere herb-gatherer for medicine. Learning in these years of disturbance was kept alive in the monasteries; and if the science of medicine was not advanced by them it was at least saved from utter oblivion.

It is quite possible that the Herbarium of Apuleius represents the amount of botanical lore that St. Augustine and his companions may have brought to England. Several manuscript translations of it into Early English—belonging to the tenth or eleventh centuries—exist at Oxford and elsewhere. In one of these, which was edited for the Rolls Series under the title of *Leechdoms, Wortcunning, and Starcraft of Early England*, many of the tenth-century English names form interesting examples of the changes wrought in Latin names by something less than five centuries of oral tradition. *Buxus* has become Box; *Cannabis*, Hænep (our Hemp); *Chærophyllum*, Cerfille (Chervil); *Malva*, Mealwe (Mallow); and *Petroselinum*, Petersilie (Parsley).

Hippocrates and Galen were read—in Latin versions, no doubt—and medicine became a serious study in which the mother-abbey of Monte Cassino took the lead. In the sixth century the neighbouring town of Salerno was the seat of a bishopric, and by the end of the seventh, of a Benedictine house. There is no doubt that some of its bishops and clergy were remarkable for learning and even for medical attainments, although it is asserted that the celebrated

School of Salerno was purely secular in its origin. An old chronicle is stated to mention a Roman named Salernus; a Jewish rabbi, Elenus; a Greek, Pontus; and a Saracen, Adala, as its joint founders at some date not stated—a story that seems to be a mere assertion of universal toleration. The School certainly claimed Charlemagne as its founder and was at first almost exclusively medical. In the ninth century the city was known as *Civitas Hippocratica*. Longfellow in his *Golden Legend* gives a good description of what was then the notion of a medical training:

> " The first three years of the college course
> Are given to Logic alone, as the source
> Of all that is noble, and wise, and true. . . .
> For none but a clever dialectician
> Can hope to become a great physician. . . .
> After this there are five years more
> Devoted wholly to medicine,
> With lectures on chirurgical lore."

The books most in vogue were, he continues:

> " Mostly, however, books of our own;
> As Gariopontus' Passionarius,
> And the writings of Matthew Platearius;
> And a volume universally known
> As the Regimen of the School of Salern,
> For Robert of Normandy written in terse
> And very elegant Latin verse."

Gariopontus's *Passionarium* was a treatise on diseases based upon Galen and written apparently at Salerno early in the eleventh century. Platearius belonged to the twelfth. William of Normandy is stated to have visited Salerno before the Conquest, and his son Robert went there to have his wounds treated on his return from the First Crusade. While there he heard of the death of his brother Rufus, and seems to have proclaimed himself, and to have been there recognised, as king of England. This explains the title *Regimen sanitatis Salerni, sive Schola Salernitanæ de conservanda bona valetudine Opusculum ad Regem Angliæ.* It also fixes the

date of its first publication as between Rufus's death in 1100
and the capture of Robert at Tenchebrai in 1106.   The
Regimen is a dietetic treatise in leonine—*i.e.* rhyming Latin
verse—and became extremely popular, being translated, and
afterwards printed, in several languages.

The later history of the School of Salerno is chiefly associated with its development into a university under the
care of that remarkable man the Hohenstaufen Emperor
Frederick II.   The Moorish foundations at Cordova have
been asserted to have been merely a group of independent
schools; but the addition of other faculties to that of medicine seems certainly to have entitled Salerno to the name of
university—perhaps the first in Europe.   Jews, and probably
Moslems also, had before this been prominent among its
scholars and teachers.

Frederick is said himself to have known the Greek, Arabic
and Hebrew languages, as well as Latin, German, and
French, and he arranged for Arab and Hebrew teachers to
be provided as well as those speaking Latin and Greek, for
the various " nations " at Salerno.   The sons of Averroes
found toleration at his court.   He learned much of Indian
plants and animals from Spanish Jews and Moorish scholars,
and encouraged the translation of the Arabic medical
writers into Latin.   Gerard of Cremona (1114–87), a
physician of Toledo, had already translated Avicenna, by
command of Frederick I.

The Arabic system of medicine seems to have exerted a
fatally formalising effect upon Salernan science.   As the
newer universities of Naples—capital of Southern Italy—
Bologna—the collegiate centre of the Papal States—and
Montpellier rose, Salerno declined, until it was finally
dissolved by Napoleon in 1811.   The *Regimen* contained
mostly common plants grown in Italian gardens.   The later
medical treatment at Salerno became largely dietetic; and
if the compounding of " simples " by the " confectionarius,"
or pharmacist, was developed into a fine art, one can easily

imagine the contempt that one of the dialecticians of Salerno would have for a " wretched, wrangling culler of herbs." Matthew Platearius, the twelfth-century physician of Salerno, seems to have been the original author of the work known from its opening words as " *Circa instans,*" the basis of the *Grand Herbier* printed at the beginning of the sixteenth century or earlier.

CHAPTER XIV

FROM CHARLEMAGNE TO ALBERTUS MAGNUS

IT is strange to turn from the far-reaching splendour of the crowning [1] of Charlemagne, the head of the Franks, as Emperor of the West, to the minute particularity of some of his imperial edicts or Capitularies. Under the wise tutelage of his intellectual prime minister Alcuin, he might found a university at Salerno or at Paris, or even establish schools in connection with every cathedral; but he can also prescribe the herbs and fruit trees that every gardener on his imperial estates is to cultivate. Some ninety of these are enumerated, including Roses, Fenugreek, Sage, Rue, Wormwood, Rosemary, Squills, Orris-root, Tarragon, Chicory, Parsley, Fennel, Endive, Dittany, Mustard, Mint, Tansy, Feverfew, the Opium Poppy, Marsh Mallow, Carrot, Parsnip, Chives, Onions, Leeks, Shallots, Madder, Fuller's Teazle, Coriander, Kohl-rabi, and the Caper Spurge. Apples of various kinds, of which eight are specified; Pears, Plums, Medlars, Service, Chestnut, Peach, Quince, Hazel, Almond, Mulberry, Bay, Fig, Walnut, and Cherry are ordered as trees. It is interesting to learn that " the gardener is to have Houseleek upon his house "—probably as a protection against lightning— and although Charlemagne's rule never extended over Britain, it would seem as if this last-mentioned order had, in the form of good advice, taken firm root among us.

Walafrid Strabo, a Swedish deacon in the monastery of St. Gall, addressed (also in the ninth century) to his abbot a poem called *Hortulus*, and his little garden of twenty-three plants are all included in Charlemagne's list.

In the same way that the writer of the *Herbarium* has been

[1] By Pope Leo III, on Christmas Eve, A.D. 800.

confused with an Apuleius of far greater literary rank, so
the name of a Classical poet seems to have been taken for a
very jejune twelfth-century performance that achieved considerable popularity in this dark age of botany. Æmilius
Macer, a native of Verona, who died in Asia, 16 B.C., is
quoted by Ovid. He is said to have written upon birds,
snakes, and medicinal plants. About A.D. 1130 or 1140—
or perhaps earlier—there was written a Latin poem, in
2160 inferior rhyming hexameters, entitled *Macer Floridus
de Naturis, Qualitatibus et Virtutibus Herbarum*. Its author
seems to have known some Greek, and is stated to have
been a physician named Odo. It is suggested that he
belonged to Salerno. The poem begins:

> " *Herbarum varias quivis cognoscere vires,*
> *Macer adest, disce, quo duce, doctus eris.*"
> (" Who longs to know the virtues of each plant,
> Macer can tell you whatso'er you want.")

The poem is divided into seventy-seven chapters, each
treating of a different plant. Among these are drugs of
distant origin, such as Pepper, Ginger, Cloves, Cinnamon,
Spikenard, Aloes, and Frankincense. Of those of European
origin the more interesting are Vervain, Marjoram, Thyme,
Horehound, Elecampane, Celandine, Woad, Hemlock, and
Henbane. There are numerous manuscripts of this poem
in existence in the Sloane, Bodleian, Ashmolean, and Cambridge collections. In the first-named is a translation of it
into English by John Lelamar, Master of Hertford School,
about A.D. 1373.

We may note in passing, that as Macer's name had become
familiar it was, in the early days of printing, affixed to two
undated Herbals that have nothing to do with this poem.

There is another short list of mediæval plant remedies
belonging to the twelfth century. These consist of some
seventy European plants, and the list is remarkable for the
extremely corrupt form of the German vernacular names
under which the plants are described. This is the *Physica*
of St. Hildegard, a remarkable woman of noble parentage,

who was born at Böckheleim, near Mayence, in 1099. Educated in the Benedictine house at Disibodenberg in Zweibrücken, she saw visions from her childhood. When nearly fifty she was recognised as a prophetess by St. Bernard of Clairvaux, by Pope Eugenius III, and by his successors during her lifetime. She became abbess at Rupertsberg near Bingen and died in 1179. Whilst most of her works are mystical, she left several hagiographies, and her treatment of herbs and trees as remedial agents occupies only the second and third of her four Books on Physics. In her description of remedies, derived mainly from the plants of her own land, she deals with them under their local names and of her own knowledge, and not by the mere copying of previous writers. In Germany this fact has caused these brief treatises to be ranked as the beginning of national biology and national medicine.

Although the general knowledge of plants in mediæval times was restricted among both Moslems and Christians to a very limited number of species considered mainly either as drugs or as articles of food, there was one writer at least— Albertus Magnus—who did something to revive the scientific study of plants as living beings. This had been practically at a standstill since the days of Aristotle.

Albert, Count (Graf) von Bollstädt (Plate II), was born at Lauingen in Suabia in 1193 or 1205–6. From a monastic school he went for his education to Padua, where he no doubt came into contact with the writings of Aristotle and may well have first acquired his taste for physical science. The writings of Moorish teachers and their translations from the Greek in Latin and vernacular versions had been spreading through Christian Europe from the time of the conquest of Toledo by Alphonsus VI, a century before. With them had spread the Moorish enthusiasm for the philosophy of Aristotle. This was known mainly through the medium of Avicenna, and was the more acceptable from his theistic setting. Averroes also was mainly considered as an expositor of Aristotle. Although neither Aristotle's belief in the eternity

ALBERT VON BOLLSTÄDT (?1193-1280).

facing p. 102.

Plate II.

of the material world, nor Averroes' distinctive theory of the emanation of the human from the divine intelligence, were at first suspect, the denunciation by orthodox Moslems of the tendency of Averroism to materialism seems to have spread to Christendom. Under pain of excommunication the Archbishop of Sens in his Provincial Synod (in 1209) forbade the reading of those of Aristotle's works that deal with natural philosophy, or of any commentary thereon, in Paris.

At Padua, Albert came under the influence of the Dominican Order, which was then in the first fervour of its foundation, for St. Dominic died in 1221, only two years before Albert entered the Order. Devoted from their foundation to the suppression of heresy, the Order of Preachers necessarily turned their attention largely to the philosophical bases of orthodoxy. Albert was trained in this Order at Hildesheim, Ratisbon and Strassburg and then began teaching at Cologne, where he spent so much of his life that he is sometimes spoken of as Albert of Cologne. Here it was that his most illustrious pupil, Thomas of Aquin, first studied under him. In 1245 they went to Paris, and, apparently at the close of a three years' course, when Albert became Doctor and Thomas Bachelor of Theology, they returned to share the teaching at Cologne.

In 1254 Albert became Provincial of his Order, and it was during his tenure of that office that he denounced the errors of Averroism then rife in the University of Paris. In 1260 Pope Alexander IV forced him to accept the Bishopric of Ratisbon, which, however, he was permitted to resign after three years. Ten years later we read of his preaching a Crusade throughout Germany, and almost his last work was the maintenance of the orthodoxy of his pupil Aquinas, who died in 1274. Albert himself died in 1280, and by his will bequeathed the whole of his property, held by papal dispensation apart from his Order, to his brethren at Cologne. His books he left to the library of the community; his ornaments to the sacristy, and his money and jewels to the com-

H

pletion of the choir of their church, which, he says, " I have
founded with my own money and built from its foundations."

Although Albert's literary works occupy twenty-one folio
volumes, they are less remarkable for their bulk than for
their systematic arrangement, by which, following Aristotle,
he covers the whole area of knowledge. The work of
Aristotle and of his Arabic commentators is presented in a
Latin digest, interpreted according to the teaching of the
Catholic Church. Albert has sometimes, therefore, been
contemptuously termed " the Ape of Aristotle."   He does not
hesitate to differ from his great master, however, as on the
question of the eternity of the world; and he has unquestionably enriched his text with much original illustration.
In his *De cœlo et mundo* he writes:

> " In studying nature we have not to inquire how God
> the Creator may, as He freely wills, use His creatures
> to work miracles and thereby show forth His power:
> we have rather to inquire what nature with its immanent
> causes can naturally bring to pass."

He does not hesitate to say that in question of natural
science he would prefer to follow Aristotle rather than
Augustine. He also says that in questions of medicine he
would prefer Hippocrates and Galen to Aristotle.

Humboldt, speaking of his *Liber cosmographicus de natura
locorum*, says it is:

> " a kind of physical geography. I have found in it
> observations regarding the simultaneous dependence of
> climate upon latitude and altitude, and the effect of
> different angles of incidence of the sun's rays in heating
> the earth's surface, which have greatly surprised me."

In his *De Mineralibus*, Albertus writes:

> " The aim of natural science is not simply to accept
> the statements of others, that is, what is narrated by
> people, but to investigate the causes that are at work
> in nature for themselves."

This and other similar frank appeals for experiment led Pouchet, who devoted a volume to Albertus and his experimentalism, to make three epochs. (i) Observation, among the Greeks, as represented by Aristotle. (ii) Compilation as represented by Pliny among the Romans. (iii) Experiment, represented in the Middle Ages by the Franciscan Roger Bacon and the Dominican Albertus.

The botanical work of Albertus, entitled *De Vegetabilibus*, is based upon the work of Nicholas of Damascus. This he believed to be Aristotle's, which he knew in its Latin dress, the work of the undetermined " Alfred." He follows this original very closely, as, for instance, in arguing that the plant is a living being, but with a vegetable soul, or grade of life lower than that of animals, and limited to nutrition, growth, and reproduction. He believes that Feeling, Will, Sleep, and Sex are not possessed by plants, although he is acquainted with what we call the sleep of the leaves of the Sensitive-plant, the periodical opening and closing of flowers and the appearance of sexuality in the Date-palm. Throughout the work there is much that is original. He holds that species are mutable, since cultivated plants, running wild, degenerate, while wild ones may be changed by domestication. Rye on a good soil, he thinks, changes to Wheat; but, if his statement of Birch and Aspen springing up from the humus when an Oak or Beech wood is cut down is capable of an interpretation in accord with fact, the further statement that Vines will spring from branches of Oak stuck into the ground is more hopeless.

It has been well said that Albertus evinces a remarkable instinct for morphology. He realises that thorns are stem structures, while prickles are superficial; that, as a bunch of grapes is sometimes replaced by a tendril, the tendril must be an incompletely developed bunch of grapes. The pentamerous symmetry of the flower of the Wild Rose and of the core of the Apple strikes him, and he notices the alternation of floral whorls, the relation of leaf-veins to the indentations of the margin, and the existence of various types of flower-form such as the star-shaped—which we now term actinomorphic—the campanulate, and the bird-like. His description of the fruit of the Apple, its three coats, the five-chambered core, the floral receptacle above and the seeds with testa and two hemispherical cotyledons, is far superior to anything in any earlier writer.

The seven books of which his book consists have been grouped as General (Books I–V), Special (Book VI), and Economic (Book VII).   It is true that in Book VI he follows the Moorish system of a merely alphabetical arrangement, whilst he does not describe a great many plants, nor any of very much interest. His *Arangus* is the Orange, *Arbor mirabilis* is Castor-oil, and *Arbor paradisi* the Plantain (*Musa paradisiaca*).   In one place he divides plants into Trees, Shrubs, Under-shrubs, Bushy herbs (*olus*), Herbs, Fungi and the like. A much more scientific grouping is hinted at, however, first into Leafless and Leaf-bearing Plants (approximating to the modern division into Cryptogams and Flowering Plants), the latter being again divided into Plants with a rind (Monocotyledons) and those with a tunicate or annular structure (Exogens). These last he would further subdivide into herbaceous and woody.

The following description of the Alder is an example of his original observation:

> "It is a tree that loves moist spots. Its wood is reddish, covered by a brown and rather smooth bark and yields a perfectly white ash. It grows in ring-like layers of wood: when dry it splits more readily than Pine and it can be preserved under water for centuries. The leaves of the Alder are rounded like those of the Pear, but are not so hard and are of a darker green. When young they are coated with a viscid humour, which has not got the perfume of the leaves of the Poplar. In winter, the Alder puts out catkins, as does the Nut-tree. In summer they are succeeded by black fruits, of the size of an Olive, resembling the cones of the Pine and enclosing the seeds."

There is a manuscript of this work at Basel, another at Strassburg, and two at Paris, and it was included in printed editions of Albert's works in 1517 and 1651. There is therefore not much excuse for writers on botany who have formed their opinion of Albertus entirely from a spurious work, *De virtutibus herbarum*, first printed in the fifteenth century, which is full of astrological nonsense.

As in the case of his great contemporary Roger Bacon, much fable as to his magical powers has collected round the name of Albertus. Not only is he said to have made a speaking machine in the form of a human head, but he is also alleged to have entertained William of Holland, King of the Romans, during the winter of 1259, in a garden of plants maintained in flower and fruit by artificial heat. The so-called Adonis gardens of the ancient Greeks were merely pots of Lettuce, Fennel, Wheat, and Barley forced during the summer, and although the hypocausts of Greek and Roman baths may seem likely to have suggested hothouses for plant culture, there does not appear to be any evidence of such a thing either before the time of Albertus or in his writings.

Of his unquestioned contributions to the science of botany we may say with Ernest Meyer:

> " No botanist who lived before Albert can be compared to him, unless it be Theophrastus, with whose work he was unacquainted; and after him none has painted nature in such living colours or studied it so profoundly until the time of Conrad Gesner and Cæsalpino."

While the encyclopædic character of his learning gained him from his contemporaries the title of Doctor Universalis, botanists certainly will not deny him his other contemporary title of the Great.

CHAPTER XV

OUR knowledge of the useful plants of other lands and our scientific acquaintance with the structure and functions of plants—to which the name of botany more properly belongs—have been derived from very different sources and have advanced spasmodically at very different times and rates. The latter we owe mainly to the educated men of science, who at most periods and in most countries have been very generally in the medical profession. The former has been accumulated by merchants, explorers or conquerors. Thus we have to mention the remarkable exploits of two travellers in the thirteenth and fourteenth centuries, neither of whom had any pretension to scientific knowledge.

During the thirteenth century the conquests of Jenghiz Khan, of his grandson Kublai Khan, and others of his family had spread not only over China but also over all Northern Asia and westward over Persia, Armenia, and parts of Anatolia and Russia. Thus Asia was more open to European intercourse than it had ever been before or has ever been since. Southern China in the middle of the century was still under a native dynasty ruling from Hang-chow. The Moslem line at Delhi still lingered and Southern India was still under Dravidian rule. As Venice was already largely engaged in the distribution of Indian spices, fabrics, and jewels over Europe, it was natural for the more enterprising among her sons to travel eastward in search of fortune.

In 1260 two brothers of noble birth, Nicolo and Maffeo Polo, were at Constantinople, the former having left a wife and a son Marco, about six years of age, at Venice (Plate IIIa). They journeyed to the Crimea, thence to Bokhara and

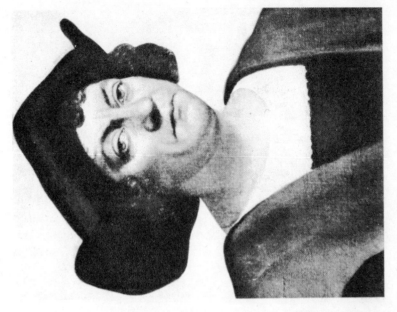

(A) MARCO POLO (? 1254-1324).

(B) CHRISTOPHER COLUMBUS (? 1451-1506).

facing p. 108.

Plate III.

ultimately to the court of Kublai Khan. The Mongol was interested in his first European visitors and sent them back as his envoys to the Pope, asking that a large body of teachers might be sent to teach his people Christianity and the Western art. They reached home in 1269, and Nicolo found that his wife had died. Two years later the two brothers once more started eastward by way of Acre, taking with them the young Marco, then a lad of seventeen, and two Dominican friars, who, however, soon turned back.

The Polos travelled by Mosul and Bagdad to Ormuz, and then, abandoning their intended sea-voyage, turned north-ward to Khorasan, Balkh, and Badakshan, up the upper Oxus to the Pamir. (This was first named in Polos' book and was not again described until visited by Lieutenant Wood in 1838.) Descending to Kashgar, Yarkand, and Khotan they reached Lob Nor (next described by Preje-valsky in 1871) and crossed the desert of Gobi, reaching Kublai's court at Shang-tu in 1275.

The young Marco rose rapidly in imperial favour and was sent on various distant missions. First he traversed Shansi and Szechuan to the borders of Tibet, Yunnan, and Burma, a region only made known to us in recent times by Baron von Richthofen. On other occasions Marco visited Kara-koram, north of the desert of Gobi, Cochin-China and Southern India. Finally, in 1292, the three Polos were sent to the court of Kublai's grand-nephew in Persia in charge of a bride for the Khan. They sailed from Chin-chew, then known as Zaitun, but were detained in Sumatra and in Southern India, so that more than two years was occupied by the journey. They afterwards reached Venice, by way of Tabriz, about 1295.

Three years later Marco Polo, while in command of a Venetian galley, was taken prisoner by the Genoese, and while in prison was persuaded by a fellow-prisoner, a Pisan named Rusticiano, to dictate an account of his travels. A year later Marco was released and he died at Venice in 1324.

Although his book seems to have been popular in his lifetime—for no less than eighty manuscripts of it exist, and it was translated into Latin before his death—he left behind him the undeserved reputation of a Munchausen. His narrative is for the most part meagre: what cosmography it has seems borrowed from Pliny, through Orosius and St. Isidore. He seems to have supplemented his own observations with his recollections of what he had learnt from the official reports of others. The fact remains, however, that to him Europe was indebted for her knowledge of China, Japan, Tibet, Burma, Siam, Cochin-China, Java, Sumatra, the Nicobar and Andaman Islands, Ceylon, and India.

Among the plants mentioned by Polo are the woods of Box in Georgia; Cotton at Mosul; the best Dates in the world at Bassora; Rice, Pomegranates, and Lemons at Ormuz; Almonds and Pistachio nuts on the Oxus; Sesame oil in Badakshan; Cotton, Linen, and Hemp in Kashgar; Bamboos a foot and a half in diameter, and Ginger on the Hoang-ho. The celebrated Chinese drug Ginseng (*Panax Ginseng* Meyer) is mentioned as well known (though not often seen by Polo himself in China), its supposed virtues being perhaps merely due to an occasional forking of the root, as in the Mandrake. Cotton, Ginger, and Sugar-cane in Bengal; Rhubarb, as we have mentioned, in Tungut; Ebony in Cochin-China; Pepper, Nutmeg, and Cloves in Borneo; Brazilwood (*Cæsalpinia Sappan* L.) in Cambodia and Sumatra; Rice, Coco-nuts, and Camphor also in Sumatra; Sandalwood in the Nicobars; the Plantain (*Musa paradisiaca* L.) in the Andaman Islands; and the Betel-Pepper (*Piper Betle* L.) in Madras.

The greatest of Moorish travellers, Ibn Batuta, was born at Tangier in 1303 and began his thirty years of travel in 1325—the year following Marco Polo's death. Much of the ground he traversed was that already visited by the Venetian, but such is the isolating effect of a hostile religion and a language so little studied in Europe as Arabic, that it was probably not until the translation of an abridgment of his narrative into English in 1829 that his work and that of

Polo were ever read by the same person. It is true that this vivacious narrative is contemporary with the so-called travels of Sir John Mandeville, but these last are now considered to be in the main fiction, while those of the Moor are in their frank self-revelation obviously genuine.

Between 1357 and 1371 this French writer—styling himself Sir Jehan de Mandeville—wrote what became, in the Middle Ages, the most popular book of travels. He has undoubtedly borrowed many of his marvels from Pliny and Solinus, and that of his alleged experiences in the Far East is borrowed from the narrative of the Franciscan Friar Odoric of Pordenone, written in 1330. He mentions the Mastic (*Pistacia Lentiscus* L.) in the Ægean and " long apples of Paradise . . . in the middle of which you will always find the figure of the Cross," presumably Plantains (*Musa paradisiaca* L.). Also some species of the same genus " whereof there are more than a hundred in a cluster," growing with " trees that bear clove-gylofres and nutmegs, and great nuts of India, and of canelle, and many other spices."

More interesting, however, is the fruit that when ripe contains " a little beast as though it were a little lamb." This is the first stage of the story by which apparently the bursting pods of the Cotton-plant, furnishing a substitute for wool, were mixed up with the hairy rhizome of a fern (*Dicksonia Baranetz* Link), the four leaf-stalks doing duty for legs, under the name of the Scythian lamb. This wonderful plant was stated to grow in the salt steppes about Samarcand; to be about two feet high, without leaves, but bearing at the top a fruit in form resembling a small woolly lamb, feeding upon the grass around it and drying up when the supply failed. It had flesh like a lobster, with blood in it, and was frequently eaten by wolves. As a matter of fact, the interior of the rhizomes of many ferns has a reddish tint. There is in the Sloane Collection at the Natural History Museum a specimen of one of these ferns that was exhibited to the Royal Society in the eighteenth century in explanation of this myth.

Ibn Batuta went from Tangier to Alexandria, Cairo, Aleppo, Damascus, Mecca, Medina, Bassora, Ispahan, Shiraz, Bagdad, Mosul, Diarbekr, and a second time to Mecca. After three years he started down the Red Sea to Aden and down the African coast to Mombasa. He then travelled north-eastward to Ormuz, across Arabia to Mecca again, and across the Red Sea to Syene and Cairo. Next he traversed Syria, Asia Minor, and the Black Sea to the Genoese colony of Caffa—now Theodosia, in the Crimea— and the Volga. He visited Constantinople, crossed the steppes to Kharism, and went by way of Bokhara and Khorasan to Cabul. He is the first Western traveller to use the name Hindu Kush for the mountains he thus crossed. He reached the Indus in September 1333, and from the coast of Sind proceeded by Multan to Delhi, where he remained eight years and was made Kazi. He was then dispatched on an embassy to the Emperor of China, and travelled through Central India to Cambay and by sea to Calicut, where for a time he was stranded by disaster. He found his way to the Maldives, however, and thence to Ceylon and Madura, returned to the Maldives, went by sea to Chittagong, inland to Silhet, and then, on a junk bound for Sumatra, to Arakan, Sumatra and thence to Zaitun (Chin-chew) and Canton. Returning and passing everywhere by means of the freemasonry of Mohammedanism, he travelled by Sumatra to the Malabar coast, to Oman, Bagdad, Damascus, Jerusalem, Cairo, Mecca—for the fourth time—and home to Fez and Tangier in 1349. After crossing to Gibraltar and making a tour through Andalusia he started once again, in 1352, for Central Africa, visited Timbuctoo and returned finally to Fez at the beginning of 1354. By order of the Sultan, the history of his seventy-five thousand miles of travel was taken down from his dictation by the royal secretary Ibn Juzai. When the French captured Constantine, in 1837, this actual fourteenth-century manuscript fell into their hands. Ibn Batuta himself died in 1378.

The plants mentioned or described by Ibn Batuta are not

very numerous and had mostly been dealt with already by
Marco Polo. The Olives and their oil, and cakes made
from the juice of the Locust (*Ceratonia Siliqua* L.) are recorded
under Syria, and Apricots and Sorghum in Arabia. He
mentions the fibrous layer of the fruit of the Coco-nut as
well as its milk and oil, and describes both the Betel-nut or
Areca palm and the Betel-pepper. The Arabic name of the
latter is Tambul, and is, he says,

> "a plant that grows like a Grape-vine. It is trained
> over a trellis of canes, like a Vine, or planted near the
> Coco-nut Palm, when it clambers up it like a Vine or a
> Pepper. It does not bear fruit; but it is the leaves
> which are used, and they are like those of the Bramble."

He distinguishes between Lemons, Citrons, and Sweet
Oranges; several times mentions Ginger and Bananas.
Under Sumatra he records Camphor, Cloves, Oranges,
Bananas, and both the Coco and the Areca Palms. It is
interesting to note that the Clove had acquired the Arabic
name *Al-garanfil*, obviously a derivative from the Greek
karyophyllon, with the article prefixed.

CHAPTER XVI

THE NEW WORLD

THE introduction of the mariner's compass into Europe at the beginning of the fourteenth century made it possible to extend maritime enterprise from the Mediterranean and the Euxine to the open Atlantic. Until then the main pathways of the human race had been the plains, the rivers and their valleys, the mountain passes or the near shore. It was Prince Henry of Portugal (1394–1460)—rightly known as the Navigator—grandson of our own John of Gaunt, who first conceived the idea of reaching the wealth of India and Arabia by the westward ocean. The discovery of the Fortunate Islands of the ancients, Porto Santo and Madeira, in 1418 and 1420, the Azores in 1432, and the Cape de Verdes in 1442, were the fruits of his zeal during his lifetime. The rounding of the Cape of Good Hope by Diaz in 1486 and the later voyages of Columbus and Vasco da Gama were also indirectly dependent upon it.

In the Canaries the Dragon-tree (*Dracæna Draco* L.), the red-wooded *Persea indica* Sp., now known as Madeira Mahogany or Vinatico, and the lichen Orchid (*Roccella tinctoria* DC.), so valuable as a dye, were discovered. At Cape Verde were found the Baobab (*Adansonia digitata* L.); and from farther south, the Malaguetta Pepper or Grains of Paradise (*Amomum Melegueta* Roscoe), that was to give its name to the Grain Coast and was used as a spice in the Elizabethan drink known as Hippocras. The Portuguese introduced into Madeira the Grape-vines of Burgundy, which were to yield the staple of the islands, and the Sugar-cane.

Meanwhile the capture of Constantinople by the Turks in 1453, an event that synchronised with the invention of printing, had scattered Greek refugee scholars over Western Europe. They brought with them the remnants of the Greek libraries, and thus caused the rapid revival of the study of the Greek language in the West. Thus it was that the dreams of Plato and Aristotle as to an Atlantis far off in the ocean, and the fortunate—though erroneous—conclusions of Ptolemy the geographer as to the small circumference of the earth, became known to many Western students. This was the *milieu* in which Christopher Columbus grew up at the University of Pavia, and during his apprenticeship to the sea (Plate III*b*).

The alleged discovery of America by Chinese Buddhist monks at the close of the fifth century has been an interesting topic for modern controversy. Bretschneider dismisses the whole story as the invention of Hoei-Sin, a " lying Buddhist and a consummate humbug," whilst Vining in his *Inglorious Columbus* (1885) maintains the truth of the entire narrative. Hoei-Sin said he had lived for some time in the kingdom of Fusang, where the "useful Mulberry-tree" furnished food, fibre, cloth, paper and timber, and that pears and grapes grew also in the land. Other distant regions were described, where the inhabitants fed on a salt plant like Wormwood, or on small beans and had clothing resembling linen. Vining maintains that Fusang is Mexico and that the Fusang tree is the *Agave*, or so-called Mexican Aloe, and the small beans Maize. The description of the tree and its uses, however, agrees better with *Broussonetia papyrifera* Vent., the Paper Mulberry of Japan, than with *Agave*. The Vine has been known in Japan from remote antiquity, and Fusang is stated by Japanese writers to be an ancient name for their country.

Though unquestionably the hardy Norseman had settled in Greenland before the close of the tenth century, and although from the twelfth to the fifteenth century there was a succession of bishops in Greenland, similar doubts attach to the accounts of Viking explorations further south. They may have visited Newfoundland and Nova Scotia—the

northern limit of the wild Vine—during the closing years of
the tenth century, bringing back to Greenland Maple wood
for fuel.   The self-sown Wheat, described in the saga of Eric
the Red, may have been *Zizania aquatica* L. the Canadian
Wild Rice, with an oat-like grain, the chief food of the
Ojibways.

Columbus knew nothing of Chinese Buddhists, however,
or of the saga of Eric the Red.   Born in Genoa, about 1436,
he was sent to the University of Pavia, where he may have
studied nautical astronomy.   Ptolemy was not printed in
Latin until 1478, and Columbus's knowledge of Aristotle's
speculations came probably from the *Imago Mundi* of Cardinal
d'Ailly (Petrus de Alliaco).   He left the university before
he was fifteen, when he decided on his career, and we know
that during the next twenty years he sailed in many seas,
visiting Iceland and the Guinea coast and at the same time,
by a profound study of the scientific bases of his art, making
himself a consummate navigator.

He reached Lisbon in 1470, and a few years later married
the daughter of one of Prince Henry's captains and the
first governor of Porto Santo.   Columbus visited that
island, earning his livelihood as a cartographer, studying his
father-in-law's log-books and pondering over the question
of undiscovered lands to the west.   In 1474 he expounded
his views to the Florentine cosmographer Paolo Toscanelli,
who was fully imbued with the views of Ptolemy that the earth
was a sphere, that its circumference was much less than it is,
and that the breadth of the then known world from the
Canaries to China was much more than it is.   He thus gave
Columbus every encouragement, sending him a map by
which the distance from the Azores westward to Cathay
was represented as only 3120 miles.

It was not until 1492 that Columbus could find a patron to
provide him with the equipment for an expedition.   He had
appealed in vain to Genoa, to Portugal—and prosecuting
the eastward route—to Henry VII of England; and his

COLUMBUS DESCRIBING HIS DISCOVERIES TO FERDINAND AND ISABELLA.

facing p. 116.

views had been denounced as heretical. At length Ferdinand and Isabella (Plate IV) agreed to his plans, and he was named Admiral and Viceroy of all lands he might discover, granted a tenth of all gain by conquest or trade, and on 3rd August, 1492, his expedition sailed in three vessels. On 16th September they arrived at the vast area of floating seaweed (*Sargassum bacciferum*) now known as the Sargasso Sea, which raised false hopes of land. On 11th October Columbus perceiving a light ahead, thought that he was approaching the island of Cipango (Japan), which Polo had said lay far east of China. On the next morning they landed on what we now know as Watling Island in the Bahamas, and, continuing the voyage, reached Cuba and Hayti.

Cotton (*Gossypium barbadense* L.), Tobacco (*Nicotiana Tabacum* L.), and Maize (*Zea Mays* L.) were among the plants that Columbus found in cultivation upon this voyage; and it is difficult to over-estimate the results of these discoveries upon the world.

Although smoking seems to have been general in the West Indies and North America and snuff-taking and chewing in the South, it is now demonstrated that no form of tobacco was so employed in the Old World. The Central American name of the plant seems to have been *Tobacco*. When it was brought to Spain in 1558, Jean Nicot, French ambassador to Portugal, sent seeds to Catherine de' Medici, and it is his name that is commemorated in that of the genus. It was not until 1586 that Drake brought tobacco from Virginia to Raleigh.

So, too, Maize was a plant unrepresented among the economic plants of the Old World. At the present day, besides its enormous production—especially as food for pigs and cattle—in the warm temperate region of the Mississippi Valley and the central United States, and in Argentina, it is largely grown in Austria, Hungary, Roumania, Russia, and India. Under the name of " mealies " it has become the staple food of a vast population in Southern Africa. In both

hemispheres it has varied under cultivation into an immense
number of forms, and is put to various other uses, such as
distillation and the manufacture of glucose.

Unlike *Nicotiana* and *Zea*, *Gossypium* was a genus of which
species were utilised in the Old World before the discovery
of America. At the present day the species have been so
hybridised that it is most difficult to unravel the pedigree of
any cultivated type. As the plant prefers moist air and dreads
frost, the largest amounts are grown in the continental areas
of the eastern United States and China, though the Soudan
promises well for the future. The best qualities are those
grown near the sea, such as the Sea Island cotton of the
south-eastern States, that of the Nile delta and the Fijis.

In his second voyage (1493–6) Columbus discovered Porto
Rico, and in his third (1498–9) Trinidad and the mouth
of the Orinoco. He died at Valladolid in 1506.

Among the plants discovered in his later voyages were the
Pine-apple (*Ananassa sativa* Lindl.), " a fruit resembling
green pine-cones with flesh like a Melon, very fragrant and
sweet "; and *Jatropha Manihot* L. (*Manihot utilissima* Pohl.),
from which the natives extracted a deadly poison but afterwards made bread. The poison of the Cassava, which plant
was speedily introduced into Tropical Africa, contains
prussic acid, whilst the granulated starch is familiar to us as
Tapioca. It is doubtful which was the greater revolution in
the distribution of economic plants, the introduction of such
American plants as Tobacco, Maize, and Potatoes to the
Old World, or the introduction of Wheat and Rice from the
Old to the New.

Humboldt sympathetically describes the effects of the
plant world around him on the mind of Columbus. He was
naturally impressed by the impenetrable jungles,

" in which one can scarcely distinguish the stems to
which the several blossoms and leaves belong. . . .
Wholly unacquainted with botany, he was led, by a
simple love of nature, to individualise all the unknown

forms he beheld. Thus, in Cuba alone, he distinguishes seven or eight different species of Palms, more beautiful and taller than the Date. He informs his learned friend Anghiera that he has seen Pines and Palms wonderfully associated together in one and the same plain; and he even observed the vegetation around him so acutely that he was the first to notice that there were Pines on the mountains of Cibao, whose fruits are not cones but berries like Olives. Thus he was the first to separate the genus *Podocarpus* from the Abietineæ."

Botanists did not arrive at this result until the work of L'Heritier at the close of the eighteenth century.

When (between 1519 and 1521) Hernando Cortes conquered Mexico, and (in 1531–3) Pizarro overthrew the Inca empire in Peru, they destroyed two portions of an extremely ancient, continuous, and highly advanced civilisation. It seems evident from Cortes' own reports and from those of Oviedo and Hernandez that the menageries and botanic gardens of Mexico, such as those of Huaxtepec, far surpassed anything then existing in Europe. Among the plants then cultivated in Mexico were Maize, Cotton, Agaves for pulque and for pita fibre, Cacao, Tomato, Capsicums, Prickly Pears, and Haricots. The Incas of Peru had the Cassava, the Banana, Maize, the Potato, Tobacco (used only as snuff), the Agave, and the Coca.

The earliest connected account of the natural history of the New World was that of Gonzalo de Oviedo (1478–1557), who lived for many years in Hispaniola as inspector of mines and alcalde. A large portion of his *Historia general y natural de las Indias* was published in 1536, illustrated by very rude engravings, among which are the earliest representations of Maize, the Pine-apple, and the Prickly Pear.

The Prickly Pear (*Opuntia Ficus-indica Miller*) was one of the first American plants introduced by the Spaniards into Europe. No less than nine varieties are described as

I

existing in Mexico at the time of the conquest, including
that upon which the cochineal insect feeds. The Moors
seem to have taken the plant to North Africa, and its use as
an effective hedge-plant in an arid climate has led to its
present wide diffusion in the tropics.

It is in Oviedo's book that we meet with the first reference
to Rubber. He describes the " Indians " as playing a
game called Batos,

> " like a game with balls, although played differently
> and the balls are of other material than those used by
> Christians."

A later writer, Antonio de Herrera (1549–1625), official
historiographer to Philip II, Philip III, and Philip IV, in his
Hiatoria General de los Hechos de los Castellanos (1601–15),
says that in the island of Hayti trees exist that when tapped
yield a milk which changes to a white gum and from which
the natives made large balls that bounced well. A little
later Juan de Torquemada, in his *De La Monarquia Indiana*
(1615), refers to the tree by the name *Ule*, which is still the
generic Mexican name for rubber plants.   He mentions that
the milk was allowed to settle in calabashes, smeared over the
bodies of the collectors to dry, or used by the Spaniards to
smear over linen coats to render them waterproof.   It was
not until De La Condamine's expedition to Peru, to measure
an arc of the meridian, that the attention of Europe was
seriously directed to the substance.   In a report in 1735
he says that

> " In the forests of the Esmeraldas grows a tree called
> *heve* by the natives; when the bark is slightly cut a
> white milk-like fluid runs out which hardens in the
> open air and becomes black," and that " the same tree
> grows on the banks of the Amazon, and its resinous fluid
> is there known as *cachuchu*. The natives make shoes of
> it which are waterproof and when smoked have the
> appearance of leather. The natives also cover moulds
> of earth, in the shape of bottles, with the material."

In a later report he says: " These bottles resemble serin-
gues " ((syringes), hence the Portuguese have called the tree
Pao de Siringa, and the Indian rubber-collectors *Seringarios*.
De La Condamine's reports led the French botanist Aublet
to investigate the trees, to which he (in 1775) gave the generic
name *Hevea*. A few years before, our own great chemist
Joseph Priestley had called attention to the utility of
caoutchouc, henceforward known as India-rubber, for rub-
bing out pencil marks. In 1823 Charles Mackintosh discovered the solubility of rubber in benzine. A little later
the invention of vulcanising, or treatment with sulphur,
and of thus making hard ebonite, led to the enormous
extension of the use of the material. In our own time the
wild produce of South American jungles is being superseded by the plantation product of south-eastern Asia, whither
Hevea brasiliensis was introduced in 1875–6 by Messrs. Cross
and Wickham, acting under the direction of the authorities
of Kew on behalf of our Indian Government.

It is to Oviedo also that we owe the first mention of the
Manioc (*Manihot utilissima*) with its poisonous starchy root,
rendered edible by washing; the Guava (*Psidium Guajava*),
the Avocado pear (*Persea gratissima*); the Calabash (*Cres-
centia cujete*); and the Sweet Potato (*Convolvulus batatas*).

Francis Lopez de Gomara (1500–60), a missionary, in his
Historia general de las Indias (1553), made known the cochi-
neal-bearing cactus (*Nopalea coccinellifera*), the Balsam of
Tolu (*Toluifera punctata*), the Agaves, sources alike of pulque
and of sisal hemp, in Mexico, and the Cacao (*Theobroma
Cacao*), long cultivated there.

The first mention of the Potato occurs in the *Cronica de
Peru* of Pedro Cieca, published at Seville in 1553. The
Spaniards had found it in cultivation near Quito, and it
seems to be a native of comparatively small altitudes in
Temperate Chile, though cultivated at great heights by the
Incas under the Equator. At the time of the Spanish con-
quests the plant was not known in Mexico. Its introduction

into Europe, which, as Prescott says, " has made an era in
the history of agriculture," is attributed to a monk named
Hieronymus Cardan, and it is also said to have been taken
from Spain to Italy by friars.   In 1585 or 1586, tubers were
apparently brought from Virginia by Drake to Raleigh's
estate near Cork, where the plant had probably been introduced less than a century before.   In 1588 the Flemish
botanist, Charles de l'Escluse—better known as Clusius—
then living at Vienna, received some from Italy, through a
member of the suite of a papal legate to Flanders.   John
Gerard, barber-surgeon, had the plant, under the name
Papus orbiculatus, in his garden in Holborn in 1596.   In
his *Herbal* in 1597 he gives an excellent figure and description
of it, and, moreover, has his own portrait on the frontispiece holding a spray of the plant (see Plate VII).   Gerard
here calls it " *Battata Virginiana sive Virginianorum, et Pappus,*
Potatoes of Virginia," and describes it as follows:

"The roote is thicke, fat and tuberous;   not much
differing either in shape, colour or taste from the
common Potatoes, saving that the rootes hereof are not
so great nor long;   some of them as round as a ball,
some small or egge-fashion, some longer and others
shorter;   which knobbie rootes are fastened unto the
stalks with an infinite number of threddie strings . . .
it hath very faire and pleasant flowers, made of one
entire whole leafe, which is folded or plaited in such
strange sort, that it seemeth to be a flower made of
sixe sundrie small leaves, which cannot easily be perceived, except the same be pulled open.   The colour
whereof it is hard to expresse.   The whole flower is of a
light purple color, stripped down the middle of every
fold or welt, with a light shew of yellownes, as though
purple and yellow were mixed togither;   in the middle
of the flower thrusteth foorth a thicke fat pointell, yellow
as golde, with a small sharpe greene pricke or point in
the middest thereof . . . it groweth naturally in
America where it was first discovered;   as reporteth C.
Clusius, since which time I have received rootes hereof

from Virginia . . . which growe and prosper in my garden, as in their owne native countrie."

As in the case of the " common " or, as we now call it, the Sweet Potato (*Convolvulus batatas*) L., it is, he continues:

" a foode as also a meate for pleasure equall in goodnesse and wholesomenesse unto the same, being either rosted in the embers, or boiled and eaten with oile, vinegar, and pepper, or dressed any other way by the hand of some cunning in cookerie."

Clusius in his *Rariorum Plantarum Historia* in 1601 called it *Papas Peruanorum*. In spite of recommendations from the Royal Society in 1663, however, it did not become popular as a food in England for another hundred years.

Jerome Benzoni, an Italian, who was in America from 1541 to 1556, in his *History of the New World* (Venice, 1556), mentions " petun "—a native name for Tobacco—and the chewing of the leaves of the Coca (*Erythroxylon Coca*) in Peru. It was the French traveller, André Thevet (1502–90), who, in his *Singularitez de la France antarctique, autrement nommée Amerique* (Paris, 1558), first describes the Ground-nut (*Arachis hypogæa*), one of the plants of America destined to have a great influence upon commerce in the Old World.

The interest aroused in Europe by the new useful and ornamental plants introduced from America is, perhaps, reflected in the fact that when Nicholas Monardes (1493–1538), a Spanish physician of Seville, published in 1569–71 an account of the West Indies—in which among other things occurs the first full account of the Tobacco—John Frampton, a London merchant who had lived some years in Seville, translated it under the title of *Joyfull Newes out of the new found world*. It is to Monardes also that our gardens are indebted for the name *Nasturtium peruanum*, afterwards translated as *Indian Cress*, for the South American plants to which Linnæus gave the name of *Tropæolum*.

In 1588 the first edition of Joseph de Acosta's *History of the Indies* appeared in Latin. The author, who was born about 1540, joined the Society of Jesus and sailed to Cartagena in 1570 as a missionary. He was afterwards stationed for many years at Juliaca on Lake Titicaca, going thence to Lima, in 1583 to Mexico, and back to Spain in 1587. He died at Salamanca in 1600. His book was quickly translated into Spanish, and later into Italian, Dutch, French, German, and English. He describes most of the American plants to which we have referred, the Prickly Pear and cochineal insect; the Brazil wood used in dyeing; the Capsicums or Chile peppers, used as a condiment and still one of the most popular esculents in the Andes; pulque, the juice of *Agave*, fermented both in Mexico and in Peru; Coca; Cacao, and the Passion-flower. The leaves of the Coca (Erythroxylon) were chewed with a little lime, as were those of the Betel Pepper in the East; but this stimulant had under Inca rule been restricted to the nobles. Its use is now almost universal among the Indians of the Andes, who will perform journeys of great length and speed with a few roasted grains of Maize, or some Beans and Coca for a digestive. From the leaves we now obtain the extremely useful anæsthetic alkaloid cocaine. Of the Cacao (*Theobroma Cacao* L.), apparently native to the hot moist basins of the Orinoco and Amazons, Acosta says that the seeds were used as money, and that the drink made from them, which bears the Mexican name of *Chocolate*, though the plant did not grow in Mexico, had been popular long before the Spanish conquest.

The Passion-flower had been discovered and named not long before by his clerical forerunners and had naturally suggested the interpretation, which, as he says, can be completed by the pious imagination. The three nail-like styles; the five slashed anthers representing the Five Wounds; the coronet, a Crown of Thorns; the ten perianth-leaves, the apostles in the absence of Judas and Peter; the spreading leaf-segments, the hands of the scoffers; and the tendrils

Botany it has too often been the practice of modern writers to ignore not only such mediæval achievement as the work of Albertus Magnus, but even the ancient greatness of Theophrastus. They have looked only at the low standard of the popular books on plants, which were the first to issue from the printing-press, in contrast to the works produced a few years later by those who have been termed the German Fathers of Botany.

The two earliest printed books dealing with plants were: *De proprietatibus rerum*, the thirteenth-century encyclopædia of the Franciscan friar Bartholomew de Glanville, and the *Buch der Natur* of Conrad of Megenburg. The former work is in nineteen or twenty books, of which the seventeenth treats " *de herbis et plantis.*" In 197 chapters it deals with 144 plants in alphabetical order, speaking chiefly of their medical uses, followed by chapters on flour, wine, oil, timber, etc., the fruit, seed, pods, roots, spines, etc. The author is especially indebted to Albertus Magnus, the Pseudo-Aristotle in the Latin version of Alfred, and Matthew Platearius of Salerno. Of numerous early editions, the first may have been printed by Caxton at Cologne in 1470.

The *Buch der Natur* merely contains an account of the medicinal virtues of a small number of plants. It is a compilation, chiefly from Pliny, St. Isidore, and Platearius, originally in Latin, but translated, perhaps by Conrad, into German in the fourteenth century. Numerous manuscripts of the work are in existence, and it ran through several editions in print in the fifteenth and sixteenth centuries.

Perhaps the earliest work dealing with plants to be printed with a definite date is a 1478 edition of the spurious astrological treatise *De virtutibus herbarum*, to which we have before referred as having so much injured the botanical reputation of Albertus Magnus. Appealing to the widespread love of the mysterious that cares little for true science, this production also went into several editions.

Of greater interest are the four early-printed editions of the *Herbarium* of Apuleius. As we have already seen, this work is of uncertain date, but had certainly been popular for centuries in manuscript. Each of the printed versions seems taken from a different manuscript, the earliest being apparently that issued in Rome by John Philip de Lignamine, physician to Pope Sixtus IV, before the close of the fifteenth century, from a manuscript found in the library at Monte Cassino. One of the chief points of interest about this very primitive herbal—a series of short chapters with names in various languages, short descriptions and notes as to properties—is that it is illustrated by figures, taken from various manuscript versions but all apparently derived from the same very ancient—possibly late Roman— originals. These symmetrical diagrammatic pictures in the manuscripts were in colour, and in the printed versions they are printed in black outline, in which, in some cases, colours have been added by stencilling. The animals for whose bites the various herbs were antidotes are also represented.

About the same time, three successive herbals made their appearance at Mayence, the cradle of the art of printing. The first was the small quarto Latin *Herbarius*, printed by Peter Schöffer in 1484 and based possibly on a manuscript of considerably earlier date. It is of unknown authorship and is compiled from Latin, Arabic, and mediæval sources. Arranged alphabetically, it was clearly intended as a popular *Materia Medica*, and was illustrated by bold outline drawings. This was speedily followed by other editions and translations issued at various places. In 1483 Schoffer issued the folio *Herbarius* in German, written or edited by Dr. Johann von Cube, a physician of Augsburg and Frankfort. This was an entirely distinct work from that of the previous year, divided into 435 chapters each dealing, in most cases, with a separate herb. It also was arranged alphabetically and illustrated by woodcuts that are remarkably better than those of the preceding publication. These woodcuts were copied and reproduced again and again, but were not improved upon until the publication of Brunfel's work in 1530.

The third work published at Mayence was the small folio *Ortus Sanitatis*, issued in 1491 by Jacob Meydenbach. It is a Latin translation of the German *Herbarius*, extended by the addition of much non-botanical matter, and copiously illustrated. Two-thirds of the woodcuts of plants are reduced copies of those in the *Herbarius*, the rest are original. The colophon ends, " Mentz, in which most noble city was first invented this most subtle art and science of character-ising or printing." Numerous editions of this work appeared during the next half-century, the French versions being named *Le Jardin de Santé*. The crudeness of the descriptions and the little acquaintance with their professed subjects shown in the woodcuts, proves the absence of any general observation of plants themselves on the part of the compilers.

Before the close of the fifteenth century several versions of another compilation similar to the *Ortus Sanitatis* were printed with illustrations copied from those in that work. This was *Le Grand Herbier*, which—translated from French into English as *The Great Herball* and " Imprynted at Lon-don in Southwarke by me Peter Treveris," in 1526—as a small folio of some 350 pages, was the first English book of its class.

The *Grand Herbier* was based seemingly upon the *Circa instans* of Matthew Platearius of Salerno. The text of the French edition ends with the statement that the book contains " the secrets of Salerno." The English version is divided into 505 chapters, in the alphabetical order of the Latin names, of which more than 400 deal with plants, each having a woodcut about two inches square, though in some cases the same figure is prefixed to different plants. Latin and English names and a lengthy account of the diseases to be cured by each plant are given, but scarcely any descriptions. William Turner, writing twenty-five years later, says the book is " al full of unlearned cacographees, and falsely naming of herbs." It was popular, however, being reprinted in 1527, 1539, and 1561.

A year before the appearance of Treveris's *Grete Herball*, there had been printed the first of a great series of popular quarto and small octavo herbals without woodcuts. This was based probably on some mediæval manuscript abridgment and each largely copied from those that precede it. The title of this first English herbal is:—" Here begynneth a newe mater, the whiche sheweth and treateth of ye vertues and proprytes of herbes, the whiche is called an Herball." As it bears no author's name but the imprint of " Rycharde Banckes, dwellynge in London, a lytel fro ye Stockes in ye Pultry," it is known as *Banckes's Herbal*. It is in quarto and contains slightly more descriptive and less medical detail than the *Grete Herball*.

Another printer, Robert Wyer, " dwellynge in saynt Martyns parysshe, at the sygne of saynt John Evangelist, besyde Charynge Crosse," published three versions of this work in octavo form. One of these in the library of Sir Joseph Banks in the British Museum has on its title-page the signature " William Shakspeare 1585." This is a forgery, however, and is perhaps the work of George Steevens. To his other editions Wyer gave the fancy name of *Macers Herbal, practysed by Doctor Lynacre*. With this the great physician Thomas Linacro very probably had as little to do as did the author of the verses known as *Macer Floridus de viribus herbarum*, which verses had been among the early products of the printing press.

Other versions of *Banckes's Herbal* appeared from the presses of T. Petit, William Myddylton, Robert Redman, William Copland, and John King. The last-named, in 1550, issued it " with certayn Additions at the ende of the boke, declaring what Herbes hath influence of certain sterres and constellations . . . by Anthony Ascham, Physycyon." Ascham, possibly a brother of the celebrated Roger, was vicar of Burniston, near Bedale, to which he was appointed by Edward VI.

Meanwhile, when the Press was thus rapidly spreading for popular consumption these products of the mediæval scriptoria, the first stage of the Renaissance was reached by the issue of Latin translations, Greek texts, and learned, if mainly textual, commentaries on the Classical writers on plants.   Pliny was printed at Venice in 1469; and, in 1492— the year before his death—the learned Venetian ecclesiastic Hermolaus Barbarus published his *Castigationes Plinianæ*.   In 1478 the same scholar had published—at Medemblik in Holland—a Latin translation of Dioscorides; and in 1499 Aldus published the original Greek text at Venice.   In 1483 Theophrastus was published in Latin by Theodore Gaza, at Treviso; and in 1495 in Greek by Aldus.   Throughout the sixteenth century botanical writers were largely concerned (their purpose being mainly pharmacological) with the identity of the drugs referred to in these Classical texts.   In the discussion of these questions there was at first but little reference to the plants themselves, nor was it realised that the species of the Eastern Mediterranean area were by no means always present in Northern Europe.

In 1536 Jean Ruel,[1] or Ruellius (1474–1537), published a folio treatise *De Natura Stirpium*, in which he quotes from all his predecessors, from Theophrastus to Hermolaus Barbarus. This work has been described as the first attempt, since the time of Theophrastus, at a general natural history of plants. Fuchs admits his indebtedness to it—though dissenting from much it contains—and considers it far in advance of any preceding work.   The arrangement is alphabetical, however, and how little its author corrected his ancient authorities by actual observation may be gauged from the fact that he insists that the Hazel has no flowers.

Another commentator on Dioscorides, of whom we would gladly know more personal details, is the Portuguese Jew

---

[1] A physician of Soissons and an excellent Greek scholar, he became a professor in the University of Paris. He devoted himself mainly to the exposition of Dioscorides, and in 1516 published a Latin translation of his master's work.

Juan Rodrigo of Castell-Branco (Johannes Rodericus de Castello Albo), who afterwards styled himself Amatus Lusitanus.[1]

Merely as explanatory of Dioscorides, such unillustrated works as those of Amatus were doomed to be superseded by those of Mattioli,[2] in which woodcuts of a high order of merit aided the descriptions. His Commentaries on Dioscorides first appeared in Italian at Venice in 1544. In 1554 it was issued in Latin as a small folio with 562 woodcuts, each about 4½ in. by 2½, of which 504 represent plants. In 1562 an edition in Bohemian in large folio was published at Prague, and a year later one in German, containing 804 larger figures of plants, the number of which, in later issues, was brought up to 1023. Unlike most of the earlier woodcuts, these are elaborately shaded and finished, and have in many instances a crowded effect.

Mattioli certainly received much help from others. Luca Ghini, the great botanical teacher of Bologna, is said to have abandoned his intention of preparing a similar work and to have given his material to Mattioli. It was mainly that Mattioli might consult it that Busbecq persuaded the Emperor to buy the Anician Codex of Dioscorides. Among the most original and interesting parts of Mattioli's work

[1] Born in 1511 and educated for the medical profession at Salamanca, he practised there and at Lisbon and travelled through France and the Netherlands. He published an Index to Dioscorides at Antwerp in 1536; lived six years at Ferrara and, in 1549, became physician to the King of Poland and later State physician at Ragusa. In 1553 he published a commentary on Dioscorides at Venice; and seems later to have lived at Saloniki and to have published various other works, including a translation of part of Avicenna from Hebrew into Latin. The date and place of his death are unknown.

[2] Pierandrea Mattioli was born at Siena in 1501, passed his childhood at Venice, where his father practised medicine, and, being intended for the legal profession, went to the University of Padua. He was soon drawn to medicine, however, which he practised in his native town, from 1521 to 1527 in Rome, and for many years at Görz. About 1554 he was called to Prague by the Emperor Ferdinand I as physician to the Archduke Ferdinand, and he occupied the same post under Maximilian II, who became Emperor in 1564; but in 1577, during an epidemic of plague, he was carried off by the disease at Trent.

are, perhaps, the notes on the plants of Asia Minor sent him by William Quakelbeen, attached as physician to Busbecq's embassy. Over 30,000 copies of the various editions of his work are said to have been sold during the lifetime of Mattioli.

No one nationality had any monopoly in the illumination of the Renaissance. For centuries the trade-routes along the Rhine had linked Flanders and the Netherlands in culture with Lyons, Marseilles, Genoa, and Venice. Now, as the printer's art spread from its birthplace at Mayence through the prince-bishoprics of the " Priests' Lane," the New Learning in all its phases, the revived study of Greek and of science and a questioning of time-hallowed dogmas of Faith, extended over the same region. Learned physicians in German-speaking lands felt, as Ruel and Mattioli felt, the need of some scientific interpretation of Dioscorides, and of some illustrations of medicinal plants better fitted to supplement their descriptions than those of the *Herbarius* or the *Ortus Sanitatis*. This desire actuated alike Brunfels, Fuchs, Bock, and both Euricius and Valerius Cordus, of whom German historians delight to speak as the Fathers of Botany, usurping a title that belongs to Theophrastus alone.

An enterprising Strassburg bookseller, John Schott— aware, perhaps, of the great skill of the artist Hans Weydiz, whose services he could secure—commissioned Brunfels,[1] an accomplished scholar, to prepare a herbal. The result was his *Herbarium vivæ eicones*—" Living Pictures of Herbs "— published in Latin, in three folio volumes, in 1530, 1531, and 1536. A German version began to appear in 1532,

---

[1] Otto Brunfels was born at Mayence in 1464, his family taking their name from Castle Brunfels near by. He entered the Carthusian monastery in his native town and remained in it until 1517, when he embraced Lutheran opinions. He became a preacher, a writer of commentaries on Holy Scripture, and—when his voice failed—a school-master. He settled at Strassburg, is said to have been for nine years head of a grammar-school there, turned his attention to medicine, obtained his doctor's degree at Basel about 1530, and enjoyed a considerable practice.

under the title of *Contrafayt Kreuterbuch* ; but in 1533 Brunfels
accepted the post of town-physician at Berne, and there he
died in 1534. A second part of the *Kreuterbuch* appeared in
1537, both Latin and German versions being published by
Schott at Strassburg.

The entirely untrue statement that Brunfels began a new
epoch in botanical study by taking his descriptions from
nature and not from books has been largely copied from
Julius von Sachs by modern writers. Brunfels' own words
are :

> " In this whole work I have no other end in view
> than that of giving a prop to fallen botany, to bring
> back to life a science almost extinct; and because this
> has seemed to me to be only possible by putting aside
> all the old herbals and publishing new and really life-
> like engravings, together with accurate descriptions
> extracted from ancient and trustworthy authors, I have
> attempted both."

The descriptions are accordingly all borrowed from books
and are very meagre. The arrangement is almost hap-
hazard, the few natural sequences being mostly adopted
from Dioscorides, though Brunfels does not fall back upon
the merely alphabetical arrangement of the Arabs. There
is a complete absence of any general notions of morphology,
and the purpose of the work is distinctly medical. At the
same time Brunfels practically anticipates by more than
two centuries the Linnæan maxim that the name of a plant
should consist of two words only.

When we turn from Brunfels' text to Hans Weydiz's 238
illustrations, however, we see that the *Herbarum eicones* does
mark a new epoch in plant iconography. Leonardo da
Vinci and Albert Dürer, for whose work that of Weydiz
has sometimes been mistaken, had both of them, before
the date of Brunfels' book, done much in the faithful reproduction of plant form by the study of the plants themselves.
But no one before Weydiz had carried out such naturalism
in the illustration of a botanical work. The clear outline

drawings—Weydiz being probably both draughtsman and engraver—are entirely different in character from the slightly later ones that appear in Mattioli's work. On the other hand, they seem to have been the suggestion for the grand work of Füllmaurer, Meyer, and Speckle, that was to appear a few years later in Fuchs' *Historia Stirpium*.

The memory of Leonard Fuchs [1] is kept evergreen by the name of that beautiful genus *Fuchsia*, which links the floras of the Andes and New Zealand.

His interest in botany, like that of Brunfels, was primarily pharmacological and his earlier writings include translations from the Greek medical classics. His first contribution to botany is a discussion of the identification of "certain herbs and simples not rightly understood." This occupies twenty-six pages in the second volume of Brunfels' *Herbarium eicones*, published in 1531. Determined apparently to outdo his friend's work, he employed the two draughtsmen Heinrich Füllmaurer and Albert Meyer to delineate plants lifesize from nature, and the best engraver in Strassburg, Veit Rudolf Speckle, to cut them in wood. It is interesting to study the physiognomies of these three artists, and of Fuchs himself, given in his *Historia Stirpium*, and published by Michael Isingrin of Basel in 1542. There is probably no engraving of the sixteenth century that excels the portrait Speckle gives of himself. The result of the work of these artists makes this the most beautiful of all herbals.

The book consists of 343 chapters dealing with as many

[1] Born at Memmingen in Bavaria, in 1501, he entered the University of Erfurt when only eleven, becoming devoted to the Classical languages. He took his bachelor's degree before he was thirteen, and after acting for some time as assistant in a private school, entered the University of Ingoldstadt, taking his Master's degree at twenty and, turning his attention to medicine, he graduated as Doctor in 1524. For two years he practised at Munich and was then called to the Professorship of Medicine at Ingoldstadt. Soon after this he became physician to the Margrave of Brandenburg and distinguished himself by his treatment of successive epidemics of plague. Having embraced Lutheran opinions, he ceased to be acceptable at Ingoldstadt, but was called in 1535 to the newly-established Protestant University of Tübingen, where he remained until his death in 1566.

K

genera, many represented by several species. It is in large
folio and contains over 500 full-page figures, and over 900
pages in all, the arrangement being primarily the alphabetical order of the Greek names. Plants unknown to the
Greeks are placed near those with which they agree in
properties. To the first edition there was prefixed a dedication, in excellent Latin, to the Margrave of Brandenburg.
Here Fuchs acknowledges the services rendered to the
science by the critical work of scholars of other nations,
such as Hermolaus Barbarus and Ruellius, as well as by
Brunfels, Bock, and Euricius Cordus. There was also
prefixed a glossary of terms, which, however, does not
evince much morphological insight. " Umbel " includes
what we now term a corymb, though the term is explained
by reference to the Greek *skiadeion* and the Latin *umbella*,
" with which women protect their complexions from the
sun and ward off the heat." " Calyx " is restricted to per-
sistent gamosepalous types; " scape " is used in our modern
sense; and *nucamenta* is introduced for what we call the
male catkins of trees, which are vaguely recognised as
taking the place of flowers.

In 1543 Fuchs published a German version of his Herbal
under the title of *New Kreuterbuch*. In this the Glossary was
omitted, but six figures were added and the descriptions
were somewhat expanded. Two years later the figures,
without text, were issued in an octavo form, and subsequently this little manual was issued with Dutch, French,
and Spanish names. These little blocks seem to have been
acquired by Jan Van der Loe of Antwerp, who employed
them, with others, in the illustration of Dodoens' *Crüyde-
boeck* of 1554 and of Clusius's French translation of it,
published as *Histoire des plantes* in 1557. Vander Loe then
probably sold them to one of the printers of the various
editions of Henry Lyte's *Nievve Herball*, a translation of
Clusius' version.

Before his death in 1566 Fuchs had prepared descriptions

and figures of 1500 plants, for which, however, he could not find a publisher.

One noteworthy reform seems to have originated with his *Historia Stirpium*—the arrangement of the description under the headings of Name, Character, Forms, Place, etc. Whilst he often has such binominal generic names as *Vitis alba*, *Vitis Idæa* or *Plantago aquatica*, he never uses such forms for specific names. Linnæus, on the other hand, two centuries later, though his generic names are all of one word, has nearly a hundred species with such binominals as *Flos-Cuculi*, *Noli-tangere*, or *Pectex-Veneris*. Fuchs used only such generic names as were descriptive, or were derived from those of patrons of the science. His specific names are, however, often merely numerical, e.g. *primum*, *alterum*, *tertium*, etc.

Whilst it seems clear that our main indebtedness to Brunfels and Fuchs is for their illustrations rather than for their text, the credit of having been the first in modern times to go direct to Nature for a knowledge of plants—usually given to Brunfels—would seem to belong rather to Jerome Bock, otherwise Hieronymus Herbarius or Hieronymus Tragus (1498-1554), *Tragos* being Greek, and *Bock*, German for Goat.[1]

He tells us how he travelled through the country in search of plants, in the disguise of a peasant. It is pleasant to read of Brunfels, already advanced in years, walking from Strassburg to Hornbach—some sixty miles—to urge his friend to write a new herbal in the language of the country. In the first volume of Brunfels' *Herbarium eicones* (1530)

[1] He was born in 1498 at Heidesbach, near Heidelberg, and well educated, though we know not where. He was intended for the cloister, but preferred the study of medicine. In 1523 was settled at Zweibrücken, in the Palatinate, at first as a schoolmaster and later as physician to the Duke with the charge of his gardens. Here he married and became a dangerously outspoken Protestant, in spite of which fact, when his patron died in 1532, and was succeeded by a Catholic prince, Bock secured a canonry at St. Fabian's, Hornbach, a few miles from Zweibrücken.

there had been many paragraphs furnished by " Hieronymus Herbarius," Jerome the Herbalist, and to the second, in 1531, he contributed a German appendix.   Now the author of that most successful work was inviting his poor country friend to supersede the German version of his own book, which was then appearing.

In 1539 Bock's *Neu Kreuterbuch* was published by Wendel Rihel of Strassburg.   It was written in German and had no illustrations.   Bock is the first botanist to attempt descriptions that should render figures unnecessary.   In his rambles he had found many plants not described in the old books and had examined and compared them with one another, noting their seasons and conditions of growth; and this knowledge he endeavours to set forth in untechnical language. In 1546 a second edition appeared as the *Kreuterbuch*, with the addition of a third part and of 567 figures drawn by David Kandel, a young self-taught artist working under Bock's own eye.   Many of these figures are re-drawn and reduced from those of Fuchs.   In 1552 a Latin translation by David Kyber appeared as " *Hieronymi Tragi de stirpium, maxime earum, quæ in Germania nostra nascuntur, usitatis nomenclaturis etc. libri tres, Germanica primum lingua conscripti, nunc in Latinam conversi.*"   This was in quarto, with the figures as before, and has an introduction by Gesner.   Although it was not reprinted, seven editions of the German version appeared after Bock's death in 1554.   As his epitaph speaks of him as " *animæ corporisque medicus,*" it would appear that— whether qualified or ordained, or neither—he practised as both minister and physician simultaneously.

Bock scorns an alphabetical arrangement and attempts with considerable success a natural arrangement, based mainly on vegetative organs, such as square stems or opposite leaves.   He had inherited from Theophrastus and Dioscorides a few natural groups, such as Umbellifers, Thistles, Legumina, Labiates, Solanaceous, Catkin-bearing and Conebearing plants.   He brings together twenty-six Crucifers, seven genera, and twice as many species of what we now term Boraginaceæ. He makes two purely biological groups, the *Serpentariæ* or climbing plants, and *Lappæ*, those with hooked prickles on the fruit. In his closer scrutiny of the flower he recognises the two deciduous sepals or " cuticuli " of Poppies; and some six or seven types of corolla—the rose-like, the stellate, the spurred, the bell-shaped, etc. He first clearly described the stamen as made up of two parts, the " capillamentum," or filament, and the " apex " or anther. He applies the name " pistillum " to the organ " like the clapper of a bell " in the flower of *Vaccinium*, though he knows nothing of its functions. He is the first writer to record the seasons of flowering and fruiting for each plant, and notes carefully the situation and the soil of each. As examples of his method we will translate his description of the Lily-of-the-Valley, not previously described, of the female flowers of the Hazel, of the essential organs of *Lilium candidum*, and of his discovery of " fern seed."

> " From a fibre bearing a root which creeps exten-sively, somewhat like that of *Lolium*, there spring in April green shoots like those of *Asparagus*, which turn out to be nothing but a pair of green leaves closely wrapped together and borne on a common stalk. When these two begin to separate they resemble a pair of the leaves of the White Lily: at the same time there arises from between them a triangular foot-stalk adorned with five or six little globes, about the size of a Chick-pea, placed one after the other, which expands towards the end of April, into little cymbal-shaped flowers of snowy whiteness, round, hollow, serrated round the lower part, marked with a purple spot within, the whole most fragrant but of a bitter taste. After the falling away of these they are followed, at the end of June, by a coral-red fruit not unlike that of the Asparagus."

> " All kinds of *Corylus* have minute red flowers, re-sembling the short stamens of a *Crocus*, which they produce just before the unfolding of their leaves. It is

in February that those stamens, which some erroneously suppose to be the flowers, acquire their yellow colour. Theophrastus, in the sixth chapter of the third book of his *History of Plants*, mentions the true flowers of *Corylus*; but Fuchs, Ruell, and others persist in denying that *Corylus* has any flowers; which they would not have done had they ever once looked into the book of Nature on the subject."

Of the flower of *Lilium* he writes:

"Out from the bottom of it there stand up six 'apices,' which are yellow; and projecting in the midst of them a kind of thickish 'stamen,' which is green and has a triangular head, the whole being shaped rather like a walking-stick."

His dependence on his own observation and freedom from superstition is shown in this description of his investigation as to "fern seed":

"Since all writers about herbs have said that ferns produce neither flowers nor seeds, I have thought well to record, for the information of botanists, my own experiences, which prove the contrary. For four successive years I kept vigil all the night before the Feast of St. John the Baptist and always found in the early morning, before daybreak some very minute black seeds, not unlike those of the Poppy, lying on the cloths and mullein leaves that I had put under the plants beforehand. Moreover, in these experiments I made use of no magic, or conjurings, or incantations."

Though far less in bulk, the contributions to botanical science of Euricius Cordus and his son Valerius are of an importance fully comparable to those of the German fathers we have discussed.[1]

[1] Euricius Cordus was born at Siemershausen in Hesse, in 1486. He received a good collegiate education at Frankenberg, and seems to have married very young and with but scanty means. When his son Valerius was born in 1515 he seems to have determined to extend his own education so that he might train the child. Within two years he had taken his Master's degree at Erfurt and was continuing his

In 1532 the former published a Latin verse translation of Nicander's *Alexipharmaca* and *Theriaca*, and in 1534 his *Botanologicon*, a dialogue between five friends during a herborising expedition, as to the reforms needed in the science. The most useful feature in this little book of 135 pages is the insistence for the first time on the fact that the plants described by Greek and Arabic authors were not likely to be identical with those known in Central Europe.

Though this little book of Euricius Cordus was not without its effect on Fuchs, Bock, and later writers, it may well be said that its author's most important work was probably the training of his son, Valerius Cordus, who has been styled " the one botanical genius of the German Renaissance." He also was born at Siemershausen, and graduated in medicine, when sixteen, at Marburg, and at Wittemberg when twenty-three. His *Dispensatorium*, written before he was twenty, was printed by order of the magistracy of Nuremberg in 1535, being thus the earliest instance of a Government Pharmacopœia. Cordus also went to the University of Leipsic and after botanising in Thuringia and Saxony he lectured in 1540 on Dioscorides at Wittemberg, notes of his lectures being published after his death.

Meanwhile he was indefatigable as a field botanist, discovering several hundred new species—more than had Brunfels, Fuchs, and Bock combined—and writing original descriptions not only of these novelties but even of plants well known to the ancient botanists. By 1540 he had thus completed a Latin manuscript *Historia Plantarum* in four

Classical studies at Leipzig. After a short but unsuccessful attempt to start a private school in conjunction with some fellow-students, he turned to the medical profession for a livelihood. With financial assistance from a friend he proceeded to Ferrara, then the leading medical school in Europe, and there graduated M.D. in 1522. After four or five years' practice at Erfurt, he became professor of medicine in the Protestant University of Marburg; but accepted the post of City Physician at Bremen shortly before his death, which took place at that city in 1535.

books, containing 446 chapters, each devoted to one species, but which was not published until 1561. Cordus went to Italy, visiting the Universities of Padua, Ferrara, and Bologna, making the acquaintance of Gesner at Zurich and of the great teacher Luca Ghini, and then going on to Florence, Pisa, Lucca, and Rome. Everywhere he collected assiduously from the cool mountains to the malarious sea-coast in the height of summer, adding a fifth book of twenty-five new species to his manuscript *Historia*. Stricken with fever, he could scarcely reach Rome, and there he died in 1544, at the early age of twenty-nine. His *Historia*, published in 1561–3 by Conrad Gesner, suffers from the unnecessary addition of 280 figures taken from Bock's *Kreuterbuch*, and in many instances misplaced.

While Bock first wrote original and full popular descriptions, Cordus is the first to draw up technical descriptions in a systematic form. It is in his 470 descriptions that we see the proofs of his botanical genius. Every description is taken from a living plant in flower and in fruit. Careful note is taken of the duration of the plant, whether annual, biennial, or perennial, and of the form, colour, odour, and taste of each part. Sections are cut through stems, fruits, and even seeds, observations being recorded as to the number of chambers in the ovary, the lines of dehiscence, the placentation and the number of rows of seeds. Cordus notes the direction in which plants twine and, as Theophrastus had done, the order in which their flowers open. He is careful never to speak of root-leaves and distinguishes between an umbel and a corymb. He frequently makes use of Theophrastus's term " pericarp," and may, in fact, owe much of his insight into plant anatomy to a study of the works of the Father of Botany more thorough than that which his predecessors had given to them. He is the first to speak of " pollen " and to use " papilionaceous " as a general term for the flowers of *Leguminosæ*. He established more new genera than any of his predecessors; recognises

Bryonia as being Cucurbitaceous and *Lithospermum* as Bor-
aginaceous; and describes no less than fifty varieties of
cultivated Pears and thirty of Apples. *Parnassia, Adoxa,
Drosera, rotundifolia,* and *Vaccinium Oxycoccus* are among the
species first fully described by him.

CHAPTER XVIII

A TRUCULENT PROTESTANT

ALTHOUGH to-day the valley of the Wansbeck at Morpeth and the river-meads of Syon, Isleworth, are at peace, and the Deanery and beautiful Cathedral Close at Wells, and even the logomachies of Cambridge wranglers, are free from bitterness, it was far otherwise in the first half of the sixteenth century. Botanists then were not free from the rancour of theological controversy. Brunfels, Fuchs, and Bock had adopted Lutheran opinions, and William Turner, the first English botanical writer of the Renaissance, was certainly not to be outdone by anyone in the violent expression of similar views.

English scholars had not been backward in welcoming the New Learning. " Grocyn," says Mr. J. R. Green, " a fellow of New College, was perhaps the first Englishman who studied under the Greek exile Chalcondylas, and the Greek lectures that he delivered in Oxford on his return mark the opening of a new period in our history. Physical as well as literary activity awoke with the rediscovery of the teachers of Greece; and the continuous progress of English science may be dated from the day when Linacre, another Oxford student, returned from the lectures of the Florentine Politian to revive the older tradition of medicine by his translation of Galen."

Aristotle, Hippocrates, Theophrastus, and Dioscorides were no longer to be obscured by their Arabic commentators, and the routine of Salerno was to give place to the direct study of anatomy and physiology.

William Turner, who has been styled " The Father of

English Botany," was probably the son of a tanner of the same name. He was born at Morpeth, Northumberland, between 1510 and 1515, and was sent to Pembroke Hall, Cambridge, at the cost of Thomas, Lord Wentworth. He graduated B.A. in 1529–30, M.A. in 1533, and in 1531 was elected Fellow. He lived in close intimacy with Nicholas Ridley, who taught him Greek, and would, he says, " often exercise himself with me both with the bow and arrow and at handball." Ridley also provided Turner with money to distribute in charity. Ridley, Turner, Latimer, and others were in the habit of meeting at the White Horse Inn to discuss theology, and the source of the opinions they adopted were such that they gained for that tavern the nickname of " Germany."

Turner may have taken deacon's Orders in 1537, about which time he published *A comparison between the Olde learnynge & the Newe*. In 1538 he issued his first botanical work, *Libellus de re Herbaria novus in quo herbarum aliquot nomina greca, latina, & Anglica habes, una cum nominibus officinarum, in gratiam studiosæ juuentutis nunc primum in lucem editus*. This little book of twenty pages is now extremely rare and consists mainly, as its title indicates, of a list of names arranged alphabetically. It also contains a few localities, chiefly in Northumberland, and these are the earliest of such local records to be printed in England. At a later date the author complained that at Cambridge he

'could learne neuuer one Greke, nether Latin, nor English name, even amongest the Phisicions of anye herbe or tre, suche was the ignorance in simples at that tyme . . and as then had nether Fuchsius nether Matthiolus nether Tragus written of herbes in Latin."

Turner married the daughter of a Cambridge alderman and became a Gospeller, " preaching without a call " until he was imprisoned for so doing, and afterwards banished. He then went to Italy, studying medicine and botany at various places. At Bologna he studied under Luca Ghini, and it was probably there that he graduated M.D. before 1544.

Ghini, though he published nothing, appears to have been a remarkable teacher. Born near Imola in 1500, he seems to have first been teacher of botany at Padua. Appointed " Reader in Dioscorides " at Bologna in 1527, he has been termed the first Professor of Botany as apart from medicine. Among his earlier pupils were Ulysses Aldrovandi, who succeeded him, and Louis Anguillara, who became " Reader in Simples " at Padua. In conjunction with them, Ghini seems to have established the Botanical Garden at Bologna to supply " simples " for demonstration to students; and it seems probable that he was the first to suggest the formation of herbaria of dried plants mounted on paper. He died in 1556.

It is very difficult to say when the earliest botanical gardens were established. We have already alluded to ancient private or royal gardens, often mainly devoted to herbs, and almost every mediæval monastery would seem to have had such a " physic garden." Hamburg claims to have had a municipal *Apothekengarten* in 1316. Salerno had one, possibly private, established by Matthæus Sylvaticus about 1340. That at Marburg, established about 1530 by Euricius Cordus, may also have been private. The garden at Venice is said to have been established in 1533 by Gaultieri, upon a site given by the State. Francis Bonafides, Professor of Medicine, had a private garden of simples at Padua in 1533, and at his recommendation it was endowed by the Venetian Senate and garden demonstrations were started in 1540. Ghini appears to have assisted here as well as at Florence and Pisa, where Cosmo de Medici established gardens in 1544. Cæsalpinus, one of Ghini's later pupils, was appointed director of the Pisan garden. Gesner's garden at Zurich belongs to 1560, and that of Paris to 1570, it being a royal foundation. Clusius founded a garden at Leyden in 1577. There was one at Leipsic in 1580. Henri IV founded that of Montpellier in 1593.

That of Nuremberg was established by Camerarius before the close of the century. Turner had gardens of his own at Cologne, Kew, and Wells, and we have the catalogue of the plants in Gerard's garden in Holborn in 1596.

Among Ghini's later pupils were Mattioli, Cæsalpinus, John Falconer, Valerius Cordus, and Turner. It is stated that he waived his intention of publishing a commentary on Dioscorides in favour of Mattioli, to whom he made over his materials. It also has been suggested that much of the wise insight into the principles of classification shown in the writings of Cæsalpinus may be directly due to his teaching.

Of John Falconer we know little beyond the fact that he was an Englishman and that he died at Ferrara in 1547. In 1540 or 1542 he had a book of dried plants, the earliest herbarium on record, which he showed to Turner. Whilst this herbarium is not known to be now in existence, those of Aldrovandus and Cæsalpinus are, the former in sixteen volumes at Bologna and the latter at Florence. Since all these men were pupils of Ghini at about the same time, there seems much probability of the truth of Sach's statement that " he seems to have been the first who made use of dried plants for scientific purposes." Other very early herbaria, however, would seem to have had an entirely independent origin. Among these are those formed by Greault at Lyons in 1558, formerly preserved at Paris; by Ratzberger in 1559, now at Cassel; by Rauwolf, in 1573-5, at Leyden; and by the Bauhins at Basel.

There is good reason to suppose that, either at Bologna in 1542, or after Cordus's early death—when it was in the hands of Conrad Gesner at Zurich—Turner saw something of the young Hessian genius's *Historia*. It was probably in 1543 that Turner first visited Gesner at Zurich, this being the beginning of a friendship and correspondence that lasted till Gesner's death more than twenty years later. In the same year Turner settled for a time at Basel, and there he probably wrote, and certainly printed, his *Huntyng and Fyndyng out of the Romish Fox, which more than seven yeares hath bene hyd among the Bysshoppes of England, after that the Kynges Hyghnes, Henry VIII., had commanded hym to be dryven out of hys Realme.* A year later Turner was at Cologne, when he published his excellent little book on the birds known to the ancients. This he dedicated to Henry VIII, and several little polemical books, with the result that in 1546 Henry prohibited his works as containing matter contrary to his own *Necessary Doctrine for any Christian Man.* Turner says that at this period he acted as physician to " the Erle of Emden " and explored some of the Frisian islands; and that he had then already written his Herbal, but had postponed its publication because he had as yet had no opportunity of visiting the west of England.

On the accession of Edward VI, Turner returned to England, became physician to the Protector Somerset, who was then established in the former Bridgettine Convent at Syon House, Isleworth. He was given the Oxford degree of M.D. on his appointment, and mentions in his books both the ducal gardens and his own private garden at Kew, the situation of which is unknown. The tradition is not improbable that attributes to his introduction the fine old Black Mulberry trees at Syon House and the adjoining Syon Lodge. From Syon House he issued in 1548 his second botanical work, *The Names of Herbes in Greke, Latin, Englishe, Duche and Frenche wyth the commune names that Herbaries and Apotecaries vse,* dedicated to Somerset. This is practically an enlarged English version of the *Libellus* of ten years before.

He pestered Cecil for Church preferment—pleading that his " chylder have bene fed so long with hope that they are very leane "—and asked among other things to be made Provost of Oriel and President of Magdalen. At last he obtained a Prebend at York and, in 1550, the Deanery of Wells, although he was not ordained priest until 1552. Turner's acrimoniousness is, perhaps, partly excused by the fact—which appears in one of his letters to Cecil—that he was so " vexed with the stone " that he wished for royal permission to retire to Germany and " drynk only rhenishe wyne." He promised, with the encyclopædic learning of the age, to correct the New Testament and to finish his *Herball* and his books on fishes, stones, and metals. Bishop Barlow had surrendered the Bishop's Palace to Somerset, the Catholic Dean Goodman had been deprived, and the Bishop was lodged at the Deanery. This home Turner had to share, and he now complained bitterly of want of room: " I can not go to my booke for ye crying of childer & noyse yt is made in my chamber." In addition to these troubles he could not get possession of the pasturage or of the full income belonging to his office.

Then it was that the first part of his *Herball* was printed (London, 1551, folio), the second part being issued at Cologne in 1562, and the whole three parts at the same place in 1568. In the Preface to the first part he says that Somerset first set him the task, but that he had had very little leisure for field work, and that

> " there haue bene in England, and there are now also certain learned men: which haue as much knowledge in herbes, yea, and more than diverse Italianes and Germanes, whyche haue set furth in prynte Herballes and bokes of simples. I mean of Doctor Clement, Doctor Wendy, and Doctor Owen, Doctor Wotton, & Maister Falconer. Yet hath none of al these set furth any thyng."

Of Falconer we have already spoken. Edward Wotton is known as the restorer of Aristotelian zoology. Dr. John Clement—tutor to the children of Sir Thomas More, Professor of Greek at Oxford, President of the Royal College of Physicians, described by More as " the best botanist and herbalist of us all "—died, an exile for his Faith, at Malines in 1572.

On the accession of Mary, Turner had again to flee abroad, while his books were once more prohibited, a fact that accounts for their rarity. It was doubtless during this second exile that he arranged for the publication of the *Herball* at Cologne. Of the 516 woodcuts it contains, 400 are taken from the octavo edition of Fuchs, and these were probably then the property of Arnold Birckman, the Cologne printer. The *Herball* is printed in black-letter and arranged in the alphabetical order of the Latin names. While aiming chiefly at the description of the medicinal plants of Dioscorides, Turner does not spare his criticisms of Fuchs, Bock, and Mattioli for their identifications. He recognises, with Euricius Cordus, that the plants of Western Europe are by no means necessarily the species known to the ancients, but he corrects the old descriptions of both the plants and their properties by his own experience and often mentions the places where the plants grow. He is, however, mainly interested in plants as " simples " and cares nothing for affinities of structure.

On the accession of Elizabeth, Turner returned to England and was reinstated in his Deanery, and there are frequent mentions in his *Herball* of his garden at Wells. His strong dislike of bishops and of sacerdotal vestments made him a thorn in the flesh of his bishop. The following " prettie storie " appears in the Martin Mar-prelate tracts :

> " Olde Doctor Turner had a dogg full of good quallities. D. Turner having invited a B. to his table, in dinner while called his dogg, and told him that the B. did sweat (you must think he laboured hard over his trencher). The dogg flies at the B. and tooke off his corner capp (he thought belike it had bene a cheese cake) and so away goes the dogg with it to his master."

At length, in 1564, on being suspended for not wearing the surplice, Turner seems to have retired to his London house in Crutched Friars, where he died on July 7, 1568, just after the appearance of his *Herball* in its complete form.

CHAPTER XIX

FIRST INVESTIGATORS IN INDIA

THE application of the name of " Indian " to the Redskins of America and of the name " West Indies " to the central archipelago of that region constantly reminds us that Columbus probably died in the belief that he had reached Zipango, the easternmost limit of Asia. After his second daring voyage across the Atlantic, it seemed clear that there were serious difficulties in reaching the spice-yielding Indies known to Marco Polo by that route. Accordingly the King of Portugal was encouraged to follow up the achievement of Diaz by another expedition eastward.

In 1497 Vasco da Gama sailed direct for the Cape Verde Islands, and rounding the Cape against baffling south-east winds he discovered Port Natal on Christmas Day. Touching at Mozambique and Mombasa and securing an Indian pilot at Melinde, he crossed the Indian Ocean in twenty-three days to Calicut. Having secured an alliance with the native ruler—in spite of the opposition of Moslem traders, who realised the commercial meaning of his achievement—and having laden his ships with spices, he returned by the same route, having spent two years on the double journey.

The extraordinary difference in the value of Indian spices in the country of their origin and in the European market is illustrated by the fact that the profits on da Gama's cargo were 6000 per cent.! Up to that time the Sultan of Egypt had derived his main revenue from duties of five per cent. on imports and ten per cent. on exports. The whole prosperity of Venice depended upon the Indian trade, but now both Egypt and Venice recognised their common danger,

L 151

and Venice actually sent wood to Cairo to be carried across
the Isthmus to Suez to build a navy to attack the Portuguese!

In 1500 Cabral, sent to follow up da Gama's success,
accidentally found himself in Brazil, but da Gama himself in
1502 took possession of Calicut.

In 1505 Almeida went as Viceroy to Ceylon to monopolise
the cinnamon trade.   In 1510 Albuquerque captured the
valuable port of Goa—which is still Portuguese—and within
the next eleven years Malacca and the Moluccas had passed
entirely into Portuguese hands.

The noble Portuguese Fernão De Magalhaes, whom we
now know generally as Magellan, had served with Almeida
and Albuquerque. He had visited Amboyna and realised
the abundance of spices that were to be procured there,
but considering himself to have been badly treated by his
sovereign, he entered the service of Charles V.   Meanwhile
the failure of Amerigo Vespucci to find a westward outlet to
the Caribbean Sea, and the linking up of his discovery of
Venezuela with Cabral's exploration of the Brazilian coast,
made it clear to the student of geography that the most
promising westward route to the Spice Islands must be to
the south of the American continent.   Having persuaded the
Emperor that this was so, Magellan sailed from Seville in
1519 on what has been well called the most important voyage
of discovery ever made.   With a view to ascertaining whether
the desired strait might be there, he carefully explored the
estuary of the La Plata and coasted the land of the tall
Patagonians, as he named them.   He spent nearly six weeks
in penetrating the straits that now bear his name.   For four
months he sailed to the north-west across the ocean, which
he named the Pacific, to the Ladrones, and, sad to say,
perished in April 1521 in a petty fight between native chiefs
in the Philippines.

Under the command of Sebastian del Cano, the expedition
reached the Moluccas and were able to see with their own
eyes the Sago-palms, Cinnamon, Camphor, Nutmeg, and
Clove trees of that region.   They filled their ships with

spices and made the return journey round the Cape, reaching
Seville three years after their departure.   The Emperor was
naturally delighted and granted del Cano a significant coat
of arms.   With the crest of a globe, the motto " *Primus
circumdedisti me*," and for supporters two Malay kings each
holding a spice branch, the shield bore the castle of Castile,
two Cinnamon sticks in saltire between three Nutmegs and
twelve Cloves.

The seizure of Egypt in 1521 by the Turks closed the old
trade-route through Alexandria, and Portugal secured the
monopoly of the spice trade until the union of the Spanish
and Portuguese crowns in 1580.   The result of this monopoly
was that in Europe the price of Pepper, which had been
about seventeen shillings a pound in 1500, rose in 1521 to
twenty-five shillings.   The fighting between the natives
and the Portuguese—followed by that between Portuguese
and Spanish, and at a later date between Portuguese,
Dutch, French, and English—formed that struggle of which
it has been said that for few objects has more blood been
spilt than for the exclusive right to sell Cloves, whilst the
European passion for Pepper led directly to our conquest
of India.

Oviedo and Monardes had begun to make known the
products of the New World of the West; those of the Near
East were largely described by Belon; and those of India
and the Far East by Garcia da Orta.

Pierre Belon was born in 1517 at Souletiere, near Le Mans,
and studied botany under Valerius Cordus (who was only
two years his senior) at Wittenberg about 1540, accompanying him on botanical journeys through Germany and
Bohemia.   Cardinal de Tournon then provided him with
the means to take a more extensive journey.   After having
graduated in medicine at Paris, he set out in 1546 to explore
the eastern Mediterranean area, traversing Turkey, Greece,
the Ægean islands, Asia Minor, Syria, Egypt, and the
Sinaitic Peninsula, and returned in 1549.   In the gossipy
little quarto volume of his travels, published in Paris in

1553 under the title of *Les observations de plusieurs singularitez,*
etc., he discusses the animals, especially those named by
ancient authors, and also the plants, drugs, customs, arts,
and ruins in the countries he visits, and illustrates his remarks
with tolerable woodcuts.   He tells us that the Turks loved
flowers and were skilful gardeners; that they used *Smilax
aspera* and our Black Bryony (*Tamus communis*) in salads;
and that Parsley was sold in the market at Constantinople
under the name of *macédonico,* which is probably the origin
of the *macédoine* of modern cookery.   He gives a good description and figure of the Oriental Plane, the Cherry Laurel,
and the Henna.   He also describes the Cedar of Lebanon,
and it is interesting to learn that the main grove on Mount
Lebanon consisted of only twenty-eight trees at the date of
his visit.   He gives a careful account of the collection of
the balsam, known as Ladanum, from *Cistus creticus* in the
island of Crete.   In this connection it may be mentioned
that Pliny had copied from Herodotus the improbable, but
substantially true, story that it was gathered from the fleeces
of goats.   In Cyprus it was so gathered, the leaves of the
plant being viscid with the excretion, but the substance as
found in commerce contained nearly seventy per cent. of
ferruginous sand.   It used to be employed as an adulterant
of Storax, and the only variety now in commerce, that from
Crete, generally contains rosin, sand, and plumbago.

In the same year as that in which he published his Travels
Belon brought out a book *De arboribus coniferis, resiniferis,
aliis quoque nonnullis sempiterna fronde virentibus.*   Thus,
although better known for books on fish and birds, he
deserves mention among botanists.   In 1564, when only
forty-seven, he was set upon by an unknown assassin,
possibly a thieving gipsy, in the Bois de Boulogne and
murdered.

In the year previous to this tragedy a remarkable little
book had appeared dealing with the medicinal plants of
India.   This was the *Coloquios* of Garcia da Orta.   The
author, whose name appears in the Latin form as Garcias

ab Horto, in Spanish as Huerta, and on the title-page of his book as Dorta, was born at Elvas, probably before 1500. He studied at the Universities of Salamanca and Alcala de Henares. At the former, founded in the thirteenth century, there were sometimes from six to fourteen thousand students and the study of the Greek medical classics and their Arabic commentators prevailed. At Alcala, the ancient Complutum, seventeen miles from Madrid, the University—founded by Cardinal Ximenes in 1500, and rendered illustrious by his polyglot version of the Scriptures—was more devoted to Hippocrates in its medical teaching. In 1526 Garcia received the royal licence to practise, and for six years he did so at Castello de Vide, being called in 1532 to teach logic in the University of Lisbon. Early in 1534 Dom Martin Alfonso de Souza, who was apparently his feudal lord in his native province of Alemtejo, having been appointed chief admiral for India, took Garcia with him to Goa as his physician. With Dom Martin, Garcia visited the coast regions of India. In 1537 he was at the important spice-exporting port of Cochin and probably, later on, in Ceylon. He formed a large " physic garden " in Goa, in which he cultivated the edible fruits and other useful plants of India. He seems also to have had a lease of the entire island of Bombaim, now Bombay, and to have had there a garden on the site now occupied by the Victoria Gardens. Da Orta died in Goa in 1570.

His *Coloquios dos simples, e drogas he cousas medicinais da India*, printed at Goa in 1563, is a small and now very rare quarto. It was the third work in a European language printed in Asia, and its initial letters had evidently been prepared for a Catechism by St. Francis Xavier that preceded it. To it is prefixed a poem by Camoens, then resident in Goa. The book itself consists of a series of dialogues reminiscent of those in Isaak Walton's *Compleat Angler*. They deal in alphabetical order with Indian drugs from Aloes to Zedoary. Quotations from Pliny, Galen, Avicenna, Dioscorides, Ruellius, Valerius Cordus, and

Oviedo prove that Garcia had a good botanical library, and
it also appears that he possessed a museum or collection of
drugs.   Though, as a devout Catholic, somewhat contemp-
tuous of the native religion, he was a close observer of
Indian customs.   He discusses diamonds, gems, elephants,
and other curious but non-botanical topics.   He also
describes the method of using the Betel-nut, the effects of
Bhang (Hemp) and Datura poisoning, Nux-vomica, the
Mango, Mangosteen, Custard-apple, and Durian.

Europe was eager for information about the newly-
explored regions of the globe, and thus it fell to the lot of
the accomplished Flemish Latinist, Charles de l'Escluse, to
introduce to the learned world of Northern Europe many of
the new-found products both of America and of India.   He
published a précis of Garcia da Orta's work in his *Aromatum
Historia* in 1567, and this was the basis of the Italian and
French versions.   Not until 1891–5 did a satisfactory edition
of the original work—that of Count Ficalho—appear; and
only in 1913 did it obtain from Sir Clements Markham a
complete English translation.

Cristobal Acosta, of Portuguese descent, though born in
Africa, practised for some time as a surgeon in Portuguese
India.   He was also at Cochin and on the Malabar Coast,
where he became acquainted with Da Orta's work.   After
falling into the hands of corsairs and being kept for a considerable time in slavery, he settled as a physician at Burgos
in Spain.   He ultimately became a monk and died in 1582.
In 1578 he published in Spanish *Tractado de las drogas y
medicinas de las Indias Orientales*, which is largely made up of
extracts from Da Orta.   This work was also translated into
Latin by Clusius and into English by his friend James
Garret, a London apothecary who is described by Gerard
as " a curious searcher of simples."

CHAPTER XX

IN the same way that the Swiss Alps are at once the meeting-place of Italy, France, and Germany, so a small number of excellent Swiss botanists serve as a link between the students of the science in the three adjacent countries.

In the sixteenth century, France, Germany, the Low Countries, England, and Switzerland were following the earlier examples of Portugal and Spain in exploring the Far West and the Far East. They followed the example of Italy, also, in making the whole of Europe the common fatherland of learning and all their contemporaries one family in intellectual pursuits. In addition to the men we have already named, Rondelet of Montpellier, Rauwolff of Augsburg, Gesner of Zurich, and Clusius of Arras may be mentioned as representatives of different nationalities thus linked together.

Conrad Gesner, whom Linnæus styled "the ornament of his age," was born at Zurich in 1516. He was the son of a poor furrier and was early left an orphan. His education and his taste for botany he owed to his uncle, who was a Protestant preacher. In 1531 Conrad went to Strassburg, and subsequently to Bourges and Paris, to study medicine. He was led by pure love of learning, however, into a course of omnivorous reading. On his return to Zurich in 1535 he married and, although compelled to devote his whole days to teaching, he spent much of his nights in study.

He attempted to resume his medical studies at Basel, while supporting himself by compiling a Latin dictionary. In 1537, being unable to meet his expenses, however, he was glad to accept a professorship of Greek at Lausanne. Three

years later his native town gave him for the second time a
small travelling scholarship to complete his medical studies,
and he visited Montpellier. He seems only to have stayed
here for a short time and does not appear to have met the
great teacher Rondelet. He owns that he learnt some
anatomy and botany, but he was disappointed in being
unable to board with some learned physician, and so
returned to Basel, where he graduated. Directly he had
become an M.D. he was appointed to a nominal Professorship of Philosophy at Zurich. He was then able to maintain
himself by his medical practice and by tremendous literary
toil, and he made many tours among his native Alps and in
Italy, France, and Germany, studying plants and animals
and visiting libraries and scholars. In 1558 he was made
Professor of Natural History and he also became State
Physician. Although constantly employing an artist to
draw plants and animals, he himself became proficient as a
draughtsman. His collection of the dried parts of animals
systematically arranged has been described as the earliest
zoological museum. In 1560 he established the Zurich
botanical garden at his own expense.

Gesner was certainly the most learned naturalist of his
time, and has been reckoned also as among the best Greek
scholars of the age. Like Linnæus, two centuries later, he
had correspondents—such as Bock, Clusius, Turner, Rondelet, and Aldrovandi—in every land. In spite of his own
prodigious literary undertakings, he was always ready to
advise others or even to edit the works of naturalists who
had been cut off in the middle of their labours.

Though Gesner's zoological work was necessarily to a
large extent compilation, as also his work on bibliography,
he lost no opportunity for personal observation, whether it
was the fish of the Rhine or the Adriatic, or the medicinal
properties of plants. More than once he injured his health
by experiments on the latter, and on one occasion nearly
killed himself by a dose of *Doronicum*. In 1564 Zurich was
visited by an epidemic of plague, and this he combated

successfully, though with much exhaustion of his own constitution, which was never robust. In the following year the scourge returned and Gesner was stricken down in the midst of his labours. When he knew that death was upon him he begged to be carried into the museum that he had loved so well, and there he died in the arms of his wife, before he had reached the age of fifty.

Besides a host of minor publications on medicine, linguistics, and mineralogy, Gesner prepared three great works, the *Bibliotheca Universalis* (1545–9), a critical index to Latin, Greek, and Hebrew literature, the *Historia Animalium* (1551–87), and a *Historia Plantarum*. Unfortunately it is only from his letters that we can gather something of what the text of this last work would have been. He recognised species as falling into groups or genera, and as varying in minor and less constant characters. It is possible, therefore, that he had a clearer conception of classification into groups of progressively increasing generality than any of his predecessors. He also insisted that flower, fruit, and seed afford better indications of affinity than do mere habit or foliage. This sound opinion he supported by adding details of flowers and fruits to his drawings in a manner that had not been done before. Although so short-sighted as to be compelled to use spectacles, it would seem that he turned the natural imperfections of his eyes to good account.

At his death he left 1500 drawings of plants—mostly original, though some were derived from those of Fuchs and others, and 400 were already engraved. They were bequeathed to a friend who published a few in the *Life of Gesner* by J. Simler (1566), but sold the whole collection of materials to Camerarius (1534–98). Camerarius employed the blocks together with his own—without giving any indication by which they can be discriminated—in an edition of Mattioli entitled *De Plantis Epitome* (1586) and in his *Hortus Medicus* (Frankfort, 1588). After his death the blocks were used, with the addition of others by different draughtsmen, in a variety of herbals printed at Frankfort, Ulm, and

Basel. Finally, a collection of about 1000 blocks and coloured drawings, some of which are probably the work of Camerarius, were purchased about 1750 by Christoph Jacob Trew of Nuremberg, who had the drawings engraved on copper. He published them in 1751 and 1771—coloured after the originals and accompanied by such of the woodcuts as were uninjured—in two great folios, under the editorship of C. C. Schmiedel, with the title *C. Gesneri Opera Botanica per duo sæcula desiderata*. But, as Sachs says, " the work, too long delayed, remained useless to a science which had by then outstripped it."

Belonging to a later generation but by birth of the same nationality as Gesner, the elder of the two brother botanists, John and Caspar Bauhin, is slightly connected with the life-story of the great Zurich naturalist.

The Bauhins were the sons of a French doctor, a native of Amiens who had taken refuge at Basel on becoming a Protestant. Here his two sons were born, John in 1541—the year of Gesner's graduation—and Caspar nineteen years later.

John Bauhin studied first in the University of his native town, and then under Fuchs at Tubingen. About 1560 he found his way to Zurich and was taken by Gesner, who was twenty-five years his senior, on some botanising excursions in the Alps, after which Gesner described him as " *eruditissimus et ornatissimus juvenis*," and even consulted him on difficult botanical questions. Bauhin botanised in the Black Forest, Burgundy, and Lombardy and studied under Aldrovandus at Bologna. In 1561 he went to Montpellier to study anatomy under Rondelet, and after graduating in medicine was probably recommended by him as assistant to D'Alechamps in the preparation of his great *Historia plantarum Lugdunensis* (1586–7). Finding his Protestantism an obstacle to his living in Lyons, he settled in 1566 as a physician in his native city, after botanising throughout Provence and Alsace and visiting Italy. In 1570 he was appointed physician to the Duke of Würtemberg, who had a large garden at Montbeliard, and there he remained until his death in 1613.

Emulating the work that Gesner had planned, Bauhin set to work to compile from all sources an illustrated history of plants. At his death this work was unfinished. He had been assisted by his son-in-law, John Henry Cherler, a physician of Basel (who had also studied at Montpellier and become professor at Nîmes, but died before him, in 1610), and the book was ultimately issued at Yverdun, in three folio volumes, by Dominic Chabrey, in 1650–1, as *Historia Plantarum Universalis.* Though the paper and print of this edition are detestable, and the 3577 woodcuts—largely copied from Fuchs—are rough, the clearness and accuracy of the text rendered it valuable. It describes about 5000 species, giving localities where they occur and synonyms for their names, and the arrangement is as natural as it could then be made. Based upon the work of Pena and Lobel, it takes all the organs of the plant, its properties and its ecology into consideration in determining its position.

The yet more valuable work of the younger brother secured prompter, if only partial, publication. Caspar Bauhin, like his elder brother, studied at Basel, Tübingen, and Montpellier, but not under either Fuchs or Rondelet. He was also at Padua and Paris, and wherever he went he seems to have been sedulous in the study and collection of wild and cultivated plants, and in making friends among botanists. In 1581 he graduated in medicine at Basel, and in the following year was made Professor of Greek, and in 1588 of Anatomy and Botany. He was subsequently appointed City physician, Professor of Medicine and Rector of the University. He died in 1624.

Although Bauhin wrote also on anatomy, his chief botanical works formed part of a comprehensive scheme. His *Phytopinax*, published in 1596, contains terse descriptions of 2700 species, beginning with Grasses and ending with *Papilionaceæ.* Among them appears the Potato under the name which it still bears, *Solanum tuberosum.* In 1620 he published the *Prodromus Theatri Botanici*, a sample of his contemplated work. This contained admirably terse and orderly diagnoses of 600 plants that he believed to be new to science, and included excellent figures of 140 of them, including the Potato. This was followed, three years later, by the *Pinax Theatri Botanici*, a systematic synonymy to all previously described species. This, his most valuable work, was, he tells us, the result of forty years' labour. It deals with about 6000 species, each briefly diagnosed with names mostly binominal, under a systematic arrangement resembling that of Lobel. He begins with the Grasses, which he considered the simplest of flowering plants, and goes on to Lilies, Gingers, dicotyledonous herbs and shrubs, to trees.

The names preferred by Caspar Bauhin were mostly adopted by Morison, Ray, and Tournefort; and the *Pinax* is quoted throughout in Parkinson's *Theatrum* and in Linnæus's *Species Plantarum*. It is thus, as Sachs says, " still indispensable for the history of individual species—no small praise to be given to a work that is more than 250 years old."

Of the twelve folio parts of which his *Theatrum* was to consist, Bauhin is said to have finished three, which may include the two parts here mentioned. The first book of the main work was issued by the author's son, John Caspar, in 1658.

Among the numerous plants first described by Caspar Bauhin are many Grasses, the Lilac (*Syringa persica*), and *Monotropa Hypopitys*. A regret has been expressed that the two brothers did not collaborate, but it was a happy thought that led Plumier, a century later, to dedicate to their joint memory the genus *Bauhinia*. This comprises about 150 species, many of which are among the most remarkable lianes or " turtle-ladders " of the forests of the Tropics, but whose most striking character is the singular twin lobes of their leaves, thus recalling this *par nobile fratrum*.

CHAPTER XXI

THE SCHOOL OF MONTPELLIER AND FLANDERS

THE medical school that existed at Montpellier early in the twelfth century owed its foundation largely to Jewish teachers educated in the Moorish universities of Spain. The school grew in prosperity as Salerno declined, and became distinguished for the empirical or practical character of its teaching, as opposed to the dogmatic routine of the school of Hippocrates. In the middle of the sixteenth century Montpellier was attracting medical students from all parts of Europe, more especially by the teaching of Rondelet, satirised by Rabelais under the thinly-veiled name " Rondibilis."

Though apparently mainly one of the old school, an exponent of Dioscorides, William Rondelet occupied a position similar to that of his contemporary Ghini. Rondelet, the younger son of an " aromatarius "—*i.e.* grocer, druggist, and pharmacist at Montpellier, was born in 1507. Left an orphan early, he was sent by his eldest brother not only to the University of his native town but also to Paris, being intended for the cloister. Preferring medicine, however, he matriculated in the medical school at Montpellier in 1529. On taking his first degree, qualifying him for practice, he set up at Pertuis at the foot of the Provençal Alps, and vainly endeavoured to eke out the income of a scanty practice by teaching children. Later he returned to Paris to learn Greek, maintaining himself in the capital by teaching, and in 1537 graduated as M.D. at Montpellier.

In 1538 Rondelet married, and it is noteworthy that the Bishop, William Pellicier, and Cardinal de Tournon stood as sponsors to two of his children. Pellicier, a student of

Pliny, was commemorated by Lobel in the *Linaria domin Pelisserii*, *L. Pelisseriana* Miller, which still adorns the wood of Grammont, the " Sylva Gramuntia " of Clusius, Lobel, and Magnol. Cardinal de Tournon, the patron of Belon, made Rondelet his physician and took him in his train to Bayonne, Bordeaux, Antwerp, and Rome. Thus it came about that Rondelet visited Ghini.

In 1545 Rondelet was made one of the Regius professors of medicine, under which title he taught anatomy, zoology, " simples," and mineralogy. In 1556 he became Chancellor of the University, and with a large and wide-stretching medical practice and three or four hours' teaching daily he was compelled to spend part of the night in study. At the same time he never lost his love of society, music, and the theatre. He established an anatomical dissecting-room, and is said to have demonstrated medicinal plants in a small garden adjoining the University. As a botanist, he seems to have been learned rather than original, however, considering plants mainly as drugs. Although his work on fishes (1554–5) was of great value to zoology, he published nothing on botany.

Eloquent and winning as a teacher, without pedantry and with much common-sense, Rondelet came at a time when there was much danger that the cult of the rediscovered Classical authors might become a superstition—as did that of Linnæus under Sir James Edward Smith and his followers. As in the case of Socrates, Ghini, Bernard de Jussieu, and Abraham Werner, his teaching lives only in the writings of his pupils.

His first distinguished pupil was James d'Alechamps (1513–88) of Caen, who matriculated in 1545. After graduation, d'Alechamps practised medicine at Lyons and —with the help, for a short time, of John Bauhin and afterwards of John Desmoulins—compiled the two folio volumes of the *Historia generalis plantarum, Lugduni*. This work, which was (in 1587) without the author's name and is generally quoted as " *Hist. Lugd.*," describes and figures more than 2700 species. They are crudely grouped in classes mainly ecological, such as climbing, bulbous, prickly, maritime, and fragrant plants, woodland trees, and cereals. Tournefoot excuses its many faults on the ground that it was thirty years in preparation.

In 1551 there came to Montpellier one who has been called the " prince of descriptive botanists," and who forms the first link between that University and Flanders. This was Charles de Lescluze, as he signed the matriculation roll, though he afterwards wrote his name de l'Escluse and is often known as Clusius. Born at Arras in 1526, Clusius had studied law at Louvain; philosophy at Marburg, and theology, under Melanchthon, at Wittenberg. He was an accomplished Latinist before he came to Montpellier, and was thus of service to Rondelet in improving the Latin of the latter's zoological writings. Clusius boarded in Rondelet's house for three years, and it was Rondelet who first imbued him with a taste for botany that affected his subsequent career. He returned through Piedmont and Savoy, Geneva, Basle, and Cologne to Antwerp, just after the first Flemish edition of the *Cruydebœck* of Rembert Dodoens, State Physician of Malines, had issued from the press of Jan Van der Loe. His first work, the translation of this influential herbal into French, appeared in 1557.

In Flanders Botany had gone through initial stages similar to those of other lands. In the thirteenth century Jean de Saint-Amand of Hainault, Canon of Tournay and afterwards professor of medicine at Paris, had written *De viribus plantarum*, which was printed in 1609. Remacle Fusch (1500–87) of Limbourg, Canon of Liège, who had practised medicine, published in 1541, 1542, and 1544 three opuscula on the names and properties of medicinal plants. Fusch had been a pupil of Brunfels at Strasburg, and his *Plantarum omnium nomenclaturæ* (1541) gives the Greek, Latin, German, Italian, French, and Walloon names of some 350 plants in alphabetical order without comment. His *De plantis ante hac ignotis* (Venice, 1542, 60 pp. duodecimo) is more important, as it describes the plants methodically, distinguishing the wild from the cultivated, besides giving their medicinal uses and French, German, and official names. Although these works forcibly recall the contemporary early writings of Turner, they were not followed by a Herbal. Fuchs also published a brief dictionary of medical biography and several small medical works.

The title of Father of Botany in Flanders justly belongs to Dodoens, and his work had a profound influence upon the study in England.

Rembert Dodoens, born at Malines in 1517, studied medicine at Louvain and graduated when eighteen. He then travelled in France, Italy, and Germany before settling in his native town, where he was appointed State Physician, and distinguished himself by his treatment of various epidemics of plague and cholera.

With the many-sided learnings of the age, Dodoens first produced a treatise on cosmography and published his first botanical work on fruits in 1552. He was then preparing his herbal, for which he obtained the wood-blocks from the octavo edition of Fuchs and had others prepared. More than half of the 707 cuts in the black-letter *Cruydeboeck* of 1554 are Fuchs's. In the second Flemish edition, that of 1563, there are 817 figures. Several smaller works, a duodecimo *Historia frumentorum* (1565), an octavo *Florum . . . odoratarumque . . . historia* (1568), a history of purgatives (1574), and one of the vine (1580), were all incorporated with the *Cruydeboeck* in the Latin *Stirpium historia Pemptades* in 1583, which contained 1305 woodcuts, while the posthumous edition of 1616 contains 1341.

In both the original *Cruydeboeck* and in the *Pemptades* there is some attempt at classification by natural affinities, instead of the merely alphabetical arrangement previously general. In this, Dodoens, though antedating the work of Lobel or the publication of that of Gesner or d'Alechamps, was preceded by Bock. The descriptions—although modelled on those of Fuchs—are not merely translations from his work, and the addition of localities for the plants in the Netherlands made the *Pemptades* a popular local Flora for many generations.

Clusius's French translation of the *Cruydeboeck* was translated into English, with some slight additions, by Henry Lyte (1529–1607) of Lyte's Carey, Somerset, in 1578, under the title of *A Niewe Herball*. Lyte's copy of Clusius's edition with his neat annotations in red ink is in the British Museum. His edition was printed by Van der Loe at Antwerp, although published in London, and for this Dodoens seems to have provided some additional blocks and matter. This English version went into five editions, two of which were subsequent to Gerards' *Herball*, which is mainly an English translation of the *Pemptades*. It has been plausibly suggested that it was from Lyte's book that Shakespeare borrowed some of his more recondite plant-lore.

Meanwhile, as we have seen, John Bauhin had gone to Montpellier in 1561. In April 1565 a Provençal student named Pierre Pena arrived there, and was followed a month later by Matthias de Lobel from Lille. Both were under the special tutelage of Rondelet and they were probably domiciled in his house. United by a common enthusiasm for botany they became great friends, travelling together and so collaborating that it is difficult satisfactorily to disentangle their separate achievements.

Pena was the elder and was born at Jouques, near Aix. He was destined for the army, but his eldest brother, an amateur astrologer, cast his nativity and found that he was destined for great success as a student. He was accordingly sent at the age of twenty to study medicine at Paris. In 1558 he seems to have visited Antwerp, and then travelled through Germany, Tyrol, Switzerland, Savoy, Spain, Portugal, and Italy. He was at Padua in 1558 and again in 1562, at Verona (where he had a friend, the pharmacist Calceolari) in 1563, perhaps at Pisa (where he knew Cæsalpinus), and at Bologna where he met Aldrovandus. In 1564 he was with Gesner at Zurich, and at Venice with

M

Valerand Dourez (who, with Lobel, was a native of Lille), afterwards in business at Lyons, to whom John Bauhin dedicated the Brookweed (*Samolus Valerandi*).

Lobel was born in 1538 (Plate V). His name is also written De l'Obel, *Obel* being our Abele, the White Poplar. He blazons a female figure between two Poplars, with the motto *Candore et Spe*, the white under surface of the leaves typifying candour, the green upper surface, hope.

Pena and Lobel certainly botanised assiduously with Rondelet, the learned William Pellicier (afterwards Bishop), and Assatius, Rondelet's son-in-law. Their great teacher died in July 1566, so that, as they thereupon left Montpellier, Lobel's knowledge of the neighbourhood was all obtained in little more than a year. A hot journey to visit a patient during an epidemic of dysentery was fatal to Rondelet, but that he had perceived the ability of his Flemish pupil is apparent, for he bequeathed to Lobel all his botanical manuscripts.

Pena and Lobel were at La Rochelle during the autumn of 1566, apparently *en route* for England, possibly desiring the safety of a Protestant country. They brought with them their herbarium, and during the next four years seem to have collected throughout the British Isles and to have obtained many newly-introduced exotics. In 1571 their joint work, *Stirpium Adversaria nova* (*i.e.* a new note-book on plants), was printed and published by Thomas Purfoot at the sign of the Lucrece in London, being dedicated to Queen Elizabeth and the Professors at Montpellier. In the Dedications the authors speak of themselves as practising medicine, and acknowledge assistance from Turner (who died in 1568); from Dr. Thomas Penny, a travelled botanist who was also the friend of Gesner, Clusius, and Gerard, and died in 1589; from Hugh Morgan, an apothecary who had a garden near Coleman Street, London; and from the garden of Sir William Cecil, afterwards Lord Burleigh, in the Strand.

It is impossible to separate the work of the two col-

MATTHIAS DE LOBEL (1538-1616).

facing p. 168

Plate V.

laborators, of course, though we may well suppose that much of the detailed knowledge of the plants of Southern France evinced in the *Adversaria* was Pena's, as he was native of that region. The most noteworthy feature of the book is the system of classification adopted, which is far better than any arrangement of the period and is most probably the only sixteenth-century system to be employed by anyone besides its author. Though preserving in the main the ancient division into herbs and trees, it subdivides herbs according to general habit and especially leaf-characters, thus foreshadowing the division of Monocotyledons from Dicotyledons. Beginning with Grasses, considered as a simple type, with long, narrow, simple entire leaves, we find associated the Sedges, Iris, Ginger, Rushes, and *Liliiflorae*. So too among the broad, net-veined and often cut-leaved plants, *Cruciferae*, *Umbelliferae*, *Papilionaceae*, and *Labiatae* occur together, and even the shrubby *Leguminosae* are associated with the herbaceous plants. If, on account of their leaf-form, the Ferns are placed in the midst of the Dicotyledonous series, and *Drosera* is classed with them, the true nature of such exceptional types as *Ophioglossum* and *Botrychium* is recognised. The importance of classification, moreover, was fully recognised, for the Preface speaks of " Order, than which there is nothing more beautiful in heaven or in the mind of a wise man."

About 1574 Pena and Lobel seem to have gone from England to Antwerp, where the latter settled and practised for some years. Pena, however, seems to have abandoned botany for a very lucrative branch of medical practice. He became private physician to Henri III, whom he is said to have cured with " Bardane " (Burdock), and died worth more than £600,000.

It would seem that about 1572 the three chief botanists whom Flanders had produced, Dodoens, Clusius, and Lobel, may all have been frequently together in the friendly intercourse that marked all their relations with one another.

Clusius had in 1564 found his way to Augsburg, secured the friendship of the great commercial magnates, the Fuggers —the Rothschilds of the age—and accompanied two scions of the family on a tour through France, Spain, and Portugal. He visited Gibraltar, Valencia, and Lisbon, but having " an extraordinary proneness to fracture and dislocation of the limbs," as one historian puts it, he broke an arm in falling from his horse in the neighbourhood of Gibraltar. He brought back to Antwerp nearly two hundred new species, however, and a collection of excellent drawings from nature. This collection furnished the material for his *Rariorum stirpium per Hispanias Historia*, which was published by Plantin in 1576. Some wood-blocks prepared for this work were, however, first employed by the publisher to illustrate the minor works of Dodoens. In 1571 Clusius visited Paris and England, and relates that Lobel met him in Bristol and took him to St. Vincent's Rocks.

In 1573 he was summoned to Vienna by the Emperor Maximilian II, to take charge of the newly-established imperial gardens. He introduced many exotics, studied the plants of Austria and Hungary, and paid a second visit to England in 1581, when he made the acquaintance of Sir Francis Drake, who gave him much information. In the same year he broke his ankle.

Meanwhile as early as 1568, on the death of the great anatomist, Vesalius, Philip II, wishing to have another Belgian as his physician, had offered the post to Dodoens, who, however, refused it. In 1572 the botanist lost his wife, and his home at Malines was pillaged by Spanish troops during the revolt of the Netherlands against the government of Alva. Accordingly, when a similar offer came from Maximilian II, Dodoens accepted, and thus found himself in 1574 reunited to his friend Clusius at Vienna. During his sojourn there his native city suffered as much at the hands of the army of the States as it had done at those of the Spaniards, and Dodoens, having property at stake, returned—staying some time *en route* at Cologne—and settled at Antwerp. In 1582, however, he accepted a medical chair in the new University of Leyden, and there he died in 1585, two years after Plantin had published the Latin edition of his collected botanical works, the *Stirpium historiæ Pemptades*.

CHAPTER XXII

AN HISTORIC PRINTING-HOUSE : PLANTIN AND CLUSIUS

In the heart of busy Antwerp, long one of the leading ports of the world, under the shadow of the lofty and beautiful cathedral, there is to-day a quiet and almost desolate square—the ancient Marche Vendredi, or Friday Market-place. Here stands the crooked-fronted group of sixteenth-century houses in which one of the greatest of printers and publishers lived and carried on his business.

Christophe Plantin (Plate VI), born near Tours between 1514 and 1525, was early left to his own resources. He was apprenticed to a printer at Caen, where, in 1545 or 1546, he married Jeanne Riviere and for some three years was working in Paris, before setting up as a book-binder, gilder, leather-worker, and working printer in Antwerp. His dainty workmanship soon became widely known. Cayas, secretary to Philip II, wishing to send a choice jewel to his master, commissioned Plantin to make a case for it. The work was done and the craftsman started at nightfall to deliver it in person. A party of drunken masqueraders, mistaking him for someone against whom they had a grudge, threw themselves upon him and one of them ran him through the body. The unfortunate Plantin dragged himself home, more dead than alive, and the surgeons barely succeeded in saving his life. He was thenceforth unable to stoop over manual work, and accordingly abandoned book-binding for printing.

Some years later it was at Cayas's recommendation that Plantin was chosen to print and publish Philip II's polyglot Bible, and he subsequently became printer to the Holy

CHRISTOPHER PLANTIN (1514-1589).

Plate VI.

See throughout the Spanish dominions. It was not until after the sack of Antwerp, in 1576, however, that Plantin established himself at the sign of the Golden Compasses in the house that formed the nucleus of the Musée Plantin, opening into the High Street and the Friday Market-place. Here Plantin lived; here he printed and published the works of Dodoens, Clusius, and Lobel; and here, in 1589, he died.

After his death the business was continued by his son-in-law, Jean Moretus, and by his descendants for three centuries, until, in 1876, the city of Antwerp purchased the printing-house and its contents to form a museum. This, as Mrs. Arber well says, " beggars description and . . . is an infallible recipe for transporting the imagination back to the time of the Renaissance, when printing was in its first youth, and was treated with the reverence due to one of the fine arts."

The buildings, mostly erected by Plantin, stand around a rectangular courtyard. Here is the little office with its window into the shop and the printer's desk, exactly as he left it. In the tapestried reception-room stands a spinet, indicative of evening recreation. On the walls are portraits of Plantin, of his wife, and of others, by his friend Rubens, the greatest of Antwerp painters, who designed some of his frontispieces. In the proof-reader's room, corrected proof-sheets still lie on the oak table in the sunny lattice. In the typeroom are the ancient matrices and the many beautifully executed wood-blocks. In the long, low-raftered printing-room stand the old hand presses. Up in the gables is the type foundry with its furnace and tools; and in the library are many of the choicest works that issued from the Plantin press; while wage-books and accounts materially help to give an atmosphere of old-world business.

On the death of Jean van der Loe, the printer of Dodoens' Herbal, Plantin bought the wood-blocks. When Lobel, then resident in Antwerp, had prepared a supplementary volume to the *Adversaria*, Plantin bought from Henry Purfoot, of London, the wood-blocks of that book and 800 copies of it to bind up with the new work. Dodoens' *Historia frumentorum* (1566), *Historia florum* (1568), and *Historia purgantium* (1574) were the first works by any of the three friends to issue from the Plantin press. In 1576 both Clusius's *Rariorum stirpium Hispanias* and Lobel's *Plantarum Historia* were published. This last-mentioned work consists of the *Adversaria*, which was, as we have seen, partly the work of Pena, and of Lobel's *Stirpium Observationes*, which deals chiefly with cultivated plants and contains nearly 1500 figures, on a larger scale than those in the *Adversaria*.

It is in Lobel's *Observationes* that we find the first suggestion of a generalisation that, more than two centuries later, was to form an important starting-point in the science of plant-geography, viz. that the mountain plants of warmer latitudes grow at lesser altitudes further north.

By 1581, Lobel had recast the *Historia* into Flemish, and in that year it was published under the title of *Kruydtboeck*, with nearly 2200 figures. At the same time these figures, many of which had been employed also by Plantin in the works of Dodoens and Clusius, were issued separately as *Plantarum seu stirpium icones*. This work, to which an index in seven languages was added, is arranged according to Pena and Lobel's classification, and was much used by later botanists down to the time of Linnæus. Lobel dedicated his *Kruydtboeck* to William of Orange, the Stadtholder. From 1581 until the assassination of that Prince in 1584, he acted as his physician, and apparently resided at Delft. He then became City Physician in Antwerp, but a few years later seems to have finally migrated to London. His daughter married one James Coel of Highgate, and he had charge of the physic garden of Lord Zouch at Hackney. In 1592 he accompanied this patron on an embassy to the Danish Court.

In 1583 Plantin published Dodoens's *Pemptades* and Clusius's *Rariorum Stirpium per Pannoniam, Austriam . . . Historia*. This was the result of some ten years' herborisings and painstaking plant description, and it contains 364 woodcuts. Among the treasures of the Musée Plantin is the original coloured drawing of the Potato sent to Clusius at Vienna in 1588, endorsed in Plantin's own handwriting, "*Taratoufli a Philipp. de Sivry acceptum Viennae* 26 *Januarii* 1588. *Papas Perouanum Petri Ciecæ.*" Shortly after this date, Clusius found his position as a Protestant at Vienna intolerable and accordingly went to Frankfort. Here he received another drawing of the plant from his friend James Garret, the London apothecary, who translated Christobal Acosta's treatise on Indian drugs.

Whilst at Frankfürt, Clusius received a pension from William IV, Landgrave of Hesse; but, by y₁ another accident, he broke his right thigh and was compelled for a long time to walk with crutches. The only botanist of his age who did not practise medicine, he was at length called to a Chair of Botany at Leyden in 1593. There he ended his days in peace, keenly alive and at work until his death, in 1609, at the age of eighty-four. He said of himself that he rejoiced at the discovery of a new plant as much as if he had found a rich treasure; and for him was composed the epitaph:

"When Clusius knew each plant Earth's bosom yields,
He went a-simpling in the Elysian fields."

His *Rariorum plantarum historia* of 1601, published by Moretus, contains his previous works with the addition of a treatise on Fungi, the first ever devoted to the group, and of some papers by other botanists. His *Exoticorum Historiæ* of 1605 contains much zoological matter, and also the Latin abridgments of the works of Garcia da Orta, Acosta, and Monardes. In 1611 some further descriptions of plants and animals were published posthumously under the title of *Curæ posteriores*.   Amongst other things in this volume, there are woodcuts of the male and female Papaw from drawings made by Johannes Van Ufele in Brazil.

A master in the art of graphic and concise description, Clusius had little or no taste for classification.

CHAPTER XXIII

ALTHOUGH Lyte's translation of Dodoens enjoyed a considerable measure of success, at the close of the sixteenth century there was certainly room for a popular English Herbal. What was required was a work not severely technical in form or language, with English localities, English names and homely descriptions of plants and their uses. This want was supplied by Gerard's *Herball*.

Born at Nantwich in 1545, John Gerard (Plate VII) was apparently connected with the Gerards of Ince, in Lancashire. He went to school at Willaston, not far distant, a fact that may have given rise to the unlikely story that he was a " gentleman scholar " in the household of the Earl of Pembroke at Wilton. Having at an early age taken to the study of medicine, he seems to have travelled, possibly as ship's surgeon, in the Baltic, but to have been settled in London as a barber-surgeon by the time he was thirty. Here he was given charge of Lord Burleigh's gardens in the Strand and at Theobalds.

Gerard professed to be " throughly acquainted with the generall and speciall differences, names, properties & privie markes of thousands of plants & trees," and he had what must have been an extensive garden of his own, either in Fetter Lane or Holborn. In 1595 he was elected to the Court of Assistants of the Barber-Surgeons' Company, and in the following year he printed a catalogue of the plants in his garden. This was the earliest catalogue of any single garden, and it contains the Latin names in alphabetical order of over a thousand species. It is dedicated to Burleigh, whilst a second edition in 1599—published after Burleigh's death—to which English names were added, is inscribed to Sir Walter Raleigh. It ends with a certificate, dated July 1, 1599, and signed by Lobel, stating that he had seen all the plants growing in Gerard's garden. In a copy that belonged to Petiver and is now in the Botanical Department of the British Museum, however, a pen has been drawn across this and the words " *hæc esse falsissima* " have been added, apparently by Lobel himself. Relations between Gerard and Lobel, formerly friendly, had become strained during the issue of the former's *Herball* in 1597.

It seems that John Norton, the Queen's printer, had commissioned a Dr. Robert Priest of the Royal College of Physicians—of whom little is known—to translate into English Dodoens's *Pemptades* of 1583. This Priest seems to have done, but he died before the work was printed and the manuscript came into the hands of Gerard, whose knowledge of Latin seems to have been small. The arrangement of the plants described was altered to that of Pena and Lobel, and Norton procured from Frankfürt upwards of eighteen hundred wood-blocks to illustrate the book. These blocks had been prepared to illustrate the *Neue Kreuterbuch* of Jacob Theodor of Bergzabern (1520–90), better known by the latinised name Tabernaemontanus, taken from his birthplace.

Theodor had been a pupil of Brunfels and an assistant to Bock. In the intervals of his medical practice he had devoted himself to the preparation of this herbal. The illustrations, which were published separately as *Eicones plantarum* in 1590, are mostly copied from those of Fuchs, Bock, Dodoens, Mattioli, Clusius, and Lobel. Gerard's *Herball* is certainly not merely a re-arranged translation of Dodoens, for many English localities are added, and much garrulous and homely gossip about the uses of the plants commends the book to the non-botanical reader. Such names as Traveller's Joy, Wayfaring-tree, and Clown's Wound-wort, which Gerard coined, evince a very pretty fancy on the part of the Barber-surgeon of Holborn. He

London Printed by
Adam.Islip Joice Norton
and Richard Whitakers
Anno   1636.

JOHN GERARD (1545-1612).

facing p. 173.

Plate VII.

had, however, the audacity to speak of Dr. Priest's work as having perished with him.

While the *Herball* was passing through the press, James Garret told the printer that Gerard had made many blunders. Lobel was called in to revise the work, and says that he did so " in a thousand places," when Gerard would allow no further alterations, saying that the work was sufficiently correct and that " Lobel had forgotten the English language." Hence Lobel's views on the book as published were less flattering than the commendatory letter from him that appears at the beginning of it.

Among rather more than 1800 woodcuts in the *Herball* we may mention as of special interest those of " The breede of Barnakles" and "Battata Virginiana," the Potato. Gerard's description of the former is a sad blending of accurate observation and conclusions based upon mere imagination.

> " What our eies have seene," he writes, " and hands have touched, we shall declare. There is a small Ilande in Lancashire called the Pile of Foulders, wherein are found the broken pieces of old and brused ships, some whereof have beene cast ihither by shipwracke, and also the trunks or bodies with the branches of old and rotten trees, cast up there likewise: wheron is found a certain spume or froth, that in time breedeth unto certaine shels, in shape like those of the muskle, but sharper pointed; and of a whitish colour; wherein is conteined a thing in forme like a lace of silke finely woven, as it were togither, of a whitish colour; one ende whereof is fastned unto the inside of the shell, even as the fish of Oisters and Muskles are; the other ende is made fast unto the belly of a rude masse or lumpe, which in time commeth to the shape and forme of a Bird: when it is perfectly formed, the shel gapeth open, and the first thing that appeareth is the foresaid lace or string; next come the legs of the Birde hanging out; and as it groweth greater, it openeth the shell by degrees, till at length it is all come foorth, and hangeth onely by the bill; in short space after it commeth to full maturitie, and falleth into the sea,

where it gathereth feathers, and groweth to a foule, bigger than a Mallard, and lesser than a Goose."

Gerard evidently attached much importance to his part in the introduction of the Potato, since he had his portrait engraved for his title-page, with a sprig of the plant in his hand. His woodcut was the earliest published representation of *Solanum tuberosum*, and he writes of it:

> " It groweth naturally in America where it was first discovered, as reporteth C. *Clusius*, since which time I have received rootes hereof from Virginia, otherwise called Norembega, which growe and prosper in my garden, as in their owne native countrie."

The facts of the introduction of the potato would seem to be as follows:—Sir Francis Drake came home from the West Indies in 1586, bringing back from Virginia Raleigh's second party of colonists. Drake may well have brought potatoes from Carthagena and Gerard have obtained them through his patron Raleigh. The plant appears in the catalogue of his garden as *Papus orbiculatus*. Gerard may have shown the plant to his " brother apothecarie " James Garret of Lime Street, and when—as we have mentioned—Garret sent a sketch of the plant to Clusius, that great bontanist probably replied with the " report " to which Gerard refers. Clusius had had the drawing, now at the Musée Plantin, since 1588, though he published no figure of the species till 1601, and the plant had certainly been known in Spain for some years prior to 1588.

Gerard, no doubt, gained credit by his book. From a lease dated 1604 we gather that he was " Herbarist to the King," and in 1608 he was elected Master of the Barber-Surgeons' Company. He died in February 1612, and was buried in St. Andrew's, Holborn. Among those who assisted him he mentions Edward, Lord Zouch—the patron of Lobel, who had brought plants from Crete and Constantinople—James Garret, Jean Robin—who in the year in which the *Herball* was printed was appointed by Henri IV keeper of his garden at Paris—and his " servant, William

Marshall," whom he sent as surgeon to the *Hercules* to bring home new plants from the Mediterranean.

Lobel survived Gerard by four years, dying at the house of his son-in-law, at Highgate, in 1616. His later years had not been idle. In 1605, under the title *Dilucidæ Simplicium Medicamentorum explicationes, et Stirpium Adversaria*, he republished the work that Pena and he had issued thirty-five years before, with the addition of a second part. This is chiefly devoted to the description of new Grasses and other Monocotyledons, including *Yucca gloriosa*, which flowered for the first time in England in the garden of a Mr. William Coys at Stubbers, North Okington, Essex. Lobel seems to have written a short Latin description of this plant in Gerard's *Catalogus* for 1596, expressly stating that it seemed different from the Yuca of the Indians from which bread is made, Yuca being still the common Quichua name for *Manihot*. Gerard, apparently misreading Lobel's Latin, confuses the two plants, and says that the Yucca grows all over South and Central America and that Cassava bread is made from its root. In the imprimatur to this edition Lobel is styled Botanist to King James.

Lobel had also prepared another large work, apparently in more than one volume, and this was to have been called *Stirpium Illustrationes*. It was not published when he died, but in 1655 a part of it was issued by Dr. William How, with a violent attack upon Parkinson. In this he says that Lobel's volumes were completed even to the title-page and dedication, but that Parkinson, after, as he himself puts it, " purchasing his Works with my Money," had appropriated the whole.

Paul de Lobel, apparently a son of the botanist, was apothecary to Sir Theodore Mayerne, James I's Huguenot physician. He married Mayerne's sister and lived in Lime Street, being employed in 1615 to concoct the poison for Sir Thomas Overbury who was murdered in the Tower. His servant, who took the medicine to the Tower, was smuggled away to France.

Though undoubtedly successful of itself during thirty years, Gerard's *Herball* owed the prolongation of its popularity to the excellent editing it received from Thomas Johnson, a far better botanist than Gerard. Johnson was born at Selby in Yorkshire, probably early in the seventeenth century. He became an apothecary in London, having a shop and physic garden on Snow Hill in 1629; he was already a prominent member of the Society of Apothecaries, and in 1629 he published an account in Latin of early botanical excursions of the Society of Apothecaries into Kent and to Hampstead Heath. This includes a list of the plants observed in the latter and is probably our earliest English Florula. Under the title *Mercurius Botanicus*, in 1634 and 1641, he issued similar accounts of longer tours; one, of twelve days, to Oxford, Bath, Bristol, Southampton, and the Isle of Wight, the other to North Wales, Snowdon, Anglesea, and Montgomery, where the party was entertained by the celebrated Lord Herbert of Cherbury. This last tour contains our earliest records of *Meconopsis cambrica* and *Saussurea alpina*.

Johnson's chief work, however, was his revision of Gerard, first published in 1633 and reprinted in 1636. He added over 800 species and over 700 woodcuts, so that the total of the latter exceeds 2700. Even more important, however, are his numerous and scholarly corrections, which earned for the book Ray's title of " *Gerard Emaculatus*." All the editor's additions are carefully marked, and it is stated that the entire revision was completed in a single year. On the outbreak of the Civil War Johnson joined the Royalist forces, and was rewarded for his loyalty with the degree of M.D. from the University of Oxford. He became Lieutenant-Colonel under Sir Marmaduke Rawdon and, in the defence of Basing House in 1644, was shot in the shoulder during a sortie, contracted a fever and died a fortnight later.

Among those whose assistance Johnson mentions in his edition of Gerard are the elder John Tradescant of Lambeth

—the founder of what is known as the Ashmolean Museum, now at Oxford—and John Goodyer of Mapledurham, near Petersfield, who was, perhaps, the earliest English botanist who can be called " critical."

TITLE-PAGE OF PARKINSON'S FAMOUS BOOK.

CHAPTER XXIV

THE FIRST SYSTEMATIST

In the history of Botany there is a constantly recurrent danger lest the rapid influx from all parts of the world of plants, as yet unknown and demanding description, should occupy the attention of botanical workers to the exclusion of the more essential general principles that underlie physiology, anatomy, and taxonomy. It was, therefore, a fortunate circumstance for the sixteenth century that, although the greatly promising genius of Valerius Cordus was too soon removed from earth, the age produced in Andrea Cesalpino a profound thinker who was also rich in botanical knowledge.

Cesalpino, or Cæsalpinus, was born at Arezzo in Tuscany in 1519 and studied under Ghini at Pisa. Here he graduated in medicine, succeeded Aldrovandi as director of the garden at Bologna, and in 1555 became Professor of Botany. In 1567 he added to this a medical Chair, which he retained until 1592, when he was called to Rome as physician to Pope Clement VIII and Professor at the Sapienza University. He died in Rome in 1603.

An interesting evidence of his practical study of Botany remains in his herbarium, still religiously preserved at Florence. This consists of a folio volume of 260 pages, containing 768 plants, well dried and mounted, with Latin and Italian dialect names. It was made in 1563 by order of the Grand Duke Cosmo I; was lost sight of until, on a search being instituted at the suggestion of William Sherard, it was discovered in a private library in 1717, and was removed to the care of the Museum of Natural History in 1844.

Cæsalpinus was one of the few naturalists of his age who recognised the organic nature of fossils. As a physiologist, Sir Michael Foster says of him that " he appears to have grasped the important truth, hidden, it would seem, from all before him, that the heart, at its systole, discharges its contents into the aorta and pulmonary artery, and at its diastole receives blood from the vena cava and pulmonary vein."

Although Cæsalpinus thus anticipated William Harvey, his chief work was the *De Plantis* (4to; Florence, 1583), published the same year as Dodoens's *Pemptades*. This was, perhaps, the most philosophical treatise on the subject since the days of Theophrastus. It is divided into sixteen books, the last fifteen of which are descriptive, and describes more than 1500 plants. As the author so ignored his contemporaries as to give no synonymy, however, and as he has also no illustrations, it is often difficult to identify the plants. The valuable part of the work is the first book, in which is given, in thirty pages, a full methodical exposition of the whole of theoretical botany.

Cæsalpinus is intensely Aristotelian in his method of discussion, depending, perhaps, too much upon reasoning rather than upon observation. He follows ancient authority in accepting the leaves, including those of the flower, as merely protective for buds or fruits, and denies the existence of sex in plants. Curiously enough, he combines with Aristotelian reasoning a discursive style modelled on that of Pliny.

He argues that plants have only a nutritive soul without sensation or motion, and that therefore they require less food and simpler organs than animals, and yet grow and bear fruit with greater vigour, drawing their food in water from the soil by their roots. As they have some little internal heat and a system of very fine veins, this food penetrates by capillarity. The pith in Dicotyledons— which, with the surrounding cylinder of wood, suggested an analogy with the spinal cord enclosed in the vertebral column—he considers the seat of internal heat, and from it the seed is produced, whilst leaves are produced from the rind or bark of the plant. A definite geometrical relation of leaves to one another—a phyllotaxis, in short—is recognised and the floral leaves are considered as formed exogenously, as in the case of the foliage leaves, and as metamorphosed from them. As in the then unpublished suggestions of Gesner, the fruit and seed are to be taken as the basis of classification, though perhaps mainly upon purely *a priori* grounds.

> " As the final cause of plants consists in that propagation which is effected by the seed, while propagation from a shoot (vegetatively) is of a more imperfect nature, so the beauty of plants is best shown in seed-production; for in the number of parts, the forms and varieties of the seed-vessels, the fruit-bearing (reproductive) stage shows far more adornment than the unfolding of a shoot."

In spite of such a theoretic basis, Cæsalpinus had a wealth of real observation on which to base his arguments. He clearly states that many plants have two seed-leaves while in Wheat there is only one. He recognises that in some plants food is stored up for the sprouting seedling in the cotyledons themselves, whilst in some cases they serve this purpose only and never leave the seed.

It is in its taxonomy, however, that the *De Plantis* most decidedly marked an epoch in the history of Botany. Although its author was probably unacquainted with the approximately natural system propounded by Pena and Lobel, it was, perhaps, the very characteristic that secured for him the title bestowed upon him by Linnæus, of the first true systematist, that was its main defect. Recognising that there was some natural affinity among plants to which expression should be given in a system of classification, Cæsalpinus, as in the case of many of his most illustrious successors during the next two centuries, thought that this affinity might be recognised in one set of characters, viz. the supremely important characters of the fruit and seed. His system is thus the starting-point of all the artificial carpological systems down to that of Gaertner (1788). Whilst sinking Shrubs under Trees and Under-shrubs under Herbs, he retained the two ancient groups of Trees and Herbs, dividing the first into those with usually solitary seeds and those that are many-seeded. The Herbs fall into thirteen classes, of which the last is that of the *Crypto-gamia* or, as he terms them, " *Flore fructuque carentes.*" The other twelve depend on the number of the seeds, with such subsidiary characters as a dry or fleshy pericarp; an inferior or superior ovary; the number of chambers to the ovary; or such non-carpological features as the presence or absence of bulbs or of milky juice. This classification is far less natural than that of Pena and Lobel, in which Dicotyledons and Monocotyledons were mainly kept apart. Strangely enough, it was never set forth in a clear synoptical form until Linnæus included it in his *Classes Plantarum* in 1738.

Linnæus says that " Cæsalpinus dwelt alone in the house which he had built." Whilst it is true that none of his contemporaries seem to have been capable of appreciating his work, it profoundly influenced botanists of a later date. Robert Morison (1620–83) disingenuously disclaims any such influence, although it is apparent in the system that he propounded, and his annotated copy of the *De Plantis* may be seen to this day at Oxford. Tournefort owns his indebtedness to Cæsalpinus for generic distinctions; and Linnæus also frankly admits his many obligations to him. The copy of the book that once belonged to the great Swedish naturalist, now in the library of the Linnean Society, is fully annotated in his own handwriting, Linnean names being added to supply the want of synonyms.

CHAPTER XXV

" PARADISI IN SOLE PARADISUS TERRESTRIS "

IT is an interesting fact that a short story, written by a lady in 1883 and 1884, was largely influential in arousing an interest in a seventeenth-century herbarist. As a result the market price of one of his works rose by leaps and bounds, and was subsequently re-issued in a facsimile reprint. The story was Mrs. Ewing's *Mary's Meadow* and the herbarist was John Parkinson.

John Parkinson, born in 1567, was a much older man than Thomas Johnson, the editor of Gerard. He became an apothecary and had a house and garden in Long Acre. After being apothecary to King James I, he received the title of *Botanicus Regius Primarius* from his successor, on dedicating his first volume to Queen Henrietta Maria. This was published when he was sixty-two, and his second work was published eleven years later. He died in 1650, and was buried at St. Martin's-in-the-Fields.

His first and most popular book bears the punning title *Paradisi in sole Paradisus Terrestris* (*i.e.* " Park-in-sun's Earthly Paradise "), with the sub-title of

> " A Garden of all sorts of pleasant flowers, which our English ayre will permitt to be noursed up: with A Kitchen garden of all manner of herbes, rootes, and fruites, for meate or sauce, used with us, and An Orchard of all sorte of fruit-bearing Trees and Shrubbes fit for our Land; together with the right orderinge, planting, and preserving of them, and their uses and vertues." (See illustration, p. 183.)

Among the commendatory verses prefixed are some by

Johnson. The book begins by discussing the situation, soils, and "frame" or design of the garden, with its walks, its hedges of Private, Sweete Bryer, White Thorne or *Pyracantha*, its mazes, mounts, fountains, and arbours. Nearly a thousand plants are described, and 780 are figured in 109 whole-page illustrations, from wood-blocks engraved in England. These are only partly original, part being taken from the works of Lobel and Clusius, and are of no great merit. Greenhouses and stoves were then scarcely in existence, so that almost all the plants mentioned are hardy. We read of 120 Tulipas, more than 90 Daffodils, 60 Anemonies, or Windeflowers, 50 Jacinths, 50 Carnations and Gilloflowers and 40 Flower de luces or Irises. In the orchard are over 60 varieties of Plum, as many Apples and Pears, 30 Cherries and more than 20 Peaches.

Parkinson's later work, the *Theatrum Botanicum*, published in 1640, seems to have been many years in preparation. As we have seen, he had bought Lobel's manuscripts, probably at the death of the latter in 1616. Thus we may probably consider much of his work to have been written earlier than Johnson's edition of Gerard, though this was published seven years before. In the Epistle to the Reader in the *Paradisus*, a "fourth part, A Garden of Simples," had been announced as forthcoming; but the scheme had grown, and possibly the cutting of the 2600 wood-blocks, copied for the most part from those in Gerard, may have caused some of the eleven years' delay. In the prefatory address "to the Reader" in the *Theatrum* there is a sneer at "Master Johnson's agility," which had forestalled the publication. Parkinson's work is certainly more comprehensive than Johnson's, for it describes nearly 3800 plants —as against 2850—and contains over 100 more pages, though with more than 100 fewer illustrations. Nearly the whole of Bauhin's *Pinax* is incorporated in the synonymy, and the discussion of the medicinal uses of the plants is very diffuse. The classification adopted is scarcely worthy of the name, being in some sort a retrogression to that of

Dodoens or d'Alechamps and founded on uses, habits, and habitats. When amidst the remarkable display of learning —Greek and Arabic as well as Latin—we find extensive extracts from Clusius and professedly from the Spanish Acosta and Monardes and the Portuguese Garcia da Orta, we cannot help suspecting that the great apparent originality of the whole may rather have been the posthumous work of Lobel, the intimate friend of Clusius, than of Parkinson himself.

Among those whose assistance Parkinson acknowledges are John Goodyer, the elder John Tradescant, "that pain-full industrious searcher and lover of all natures varieties " . . . " my very good friend," and the Fleming, Dr. William Boel, who had been a correspondent of Clusius and had travelled and collected seeds in Germany, Spain, Portugal, and Barbary, mainly, so Parkinson says, at his expense.

A beautiful statue of Parkinson was erected in 1902 in the Palm-house at Sefton Park, Liverpool, but his work, like Johnson's edition of Gerard, has been made a botanical classic by being systematically cited by Ray. Such is the literary charm of its quaint, straightforward, nervous English that we can be sure, in the words of one of the sets of verses prefixed to the *Theatrum*, that " No night of Age shall cloude bright Parke-in-sunne."

CHAPTER XXVI

THE FIRST TERMINOLOGIST

ONE of the most pregnant sources of error in the study of plants has always been the mistaking of analogy for homology—too great an appeal to physiology, or to final causes in the interpretation of structure—and this had led Cæsalpinus into error. The best preventive of such misconceptions is to be found in purely morphological definitions. Whilst Cæsalpinus had laid the foundations of scientific taxonomy, this complement of his work in a precise terminology was provided by Jung.

Born at Lubeck in 1587, Joachim Jung became Professor of Mathematics at Giessen, a post that he occupied for five years. He then turned to the study of medicine at the University of Rostock and at Padua, where he graduated as M.D. in 1618, and it was here, no doubt, that he became acquainted with the work of Cæsalpinus. Returning to Germany he held various Chairs at Rostock, Lubeck, and Helmstadt, and in 1629 became Rector of the Johanneum, a secondary college at Hamburg, where he died in 1657. A profound student and brilliant teacher in many departments of science—philosophical, physical, and biological— he published nothing himself, but two of his pupils printed manuscripts left by him at his death. The *De plantis Doxoscopiæ Physicæ Minores*, published in 1662, is largely systematic and is all in the form of detached aphorisms or notes made from time to time. The *Isagoge Phytoscopica*, published in 1678, is a more complete system of theoretical botany, though drawn up in a very concise form in a series of propositions arranged in strict logical sequence. It is thus strictly comparable to the first part of Theophrastus's *Historia*; to the

191

little treatise of Nicholas of Damascus; or that of Albertus Magnus, which was based upon it; to the first book of Cæsalpinus's *De Plantis*; and to the treatise *De plantis in genere* with which Ray begins his *Historia Plantarum*.

The work begins by discussing the distinction between plants and animals, the plant being, he states, a living but not a sentient being. It is fixed to its substratum, from which it obtains nourishment, feeding by assimilating such nourishment, growing when the added substance exceeds the matters thrown off, and propagating itself when it produces another individual specifically like itself. Every plant consists of a descending axis, the root, and an ascending axis with its appendicular organs, the leaves, flowers, and fruit. The stem is a prismatic body with articulations whence spring the leaves, known when protuberant as *nodes* and the lengths between them as *internodes*. The leaf has distinctly organised upper or inner, and lower or outer, surfaces. Jung was the first writer to discriminate clearly between *simple* and *compound* leaves, *pinnate* and *digitate*, *paripinnate* and *imparipinnate*, *opposite* and *alternate*, and to describe the *petiole* apart from the *blade*. He terms the calyx a *perianth*, treating petaloid perianthe such as those of Lilies as "*naked flowers*." Although quite ignorant of their functions, he calls *stamens* and *styles* by the names that we still employ. The inflorescence of the *Compositæ* is correctly described as a *capitlum*, with *disk* and *ray florets*; and fruit, seed, and embryo are carefully discriminated, as they had been by Cæsalpinus. We have here, then, a very complete outline of that precise terminology that was afterwards formulated by Linnæus. What is of still greater importance, it is purely morphological, without the least suggestion of final causes.

Improving on the teaching of Cæsalpinus and his predecessors, and following Harvey's dictum of *omne vivum ex ovo*, Jung repeatedly expresses his doubt of the existence of spontaneous generation. Sachs says of him that he was "free from the genius-stifling burden which the knowledge of individual species had become." Although he discusses the theory of system, method, or taxonomy, and the principles upon which species should be named, he insists that neither taste, colour, smell, habitat, nor medicinal uses should be employed as specific characters.

Presumably only a few copies of these two opuscula were printed; they soon became scarce, but were reprinted in a collected form in 1747. Before their first publication, however, they had become known in a quarter from which they were to influence profoundly the whole world of botanical opinion.

In 1660 a manuscript copy of the *Isagoge* had reached England and had been submitted to the critical examination of John Ray, through the good offices of Samuel Hartlib. Hartlib was a brilliant and enthusiastic Prussian merchant of Polish origin but with an English mother. He had lived in London since 1628, and " knew everybody and whom everybody knew." He had busied himself with schemes for union among Protestant bodies, for the improvement of English agriculture and of education. He seems to have communicated to his friend Milton the contents of the manuscript poem of Cædmon, discovered by Junius, which was suggestive of *Paradise Lost*; and now he was doing for botany a service similar to that which he then performed for literature.

Ray quoted, with full acknowledgment, the greater part of the *Isagoge* in the introduction to his *Historia plantarum* (1686), printing the aphorisms in italics. It was undoubtedly in that work, and there only, that Linnæus became acquainted with Jung's logical treatment of the subject.

CHAPTER XXVII

SOME EARLY DAYS AT THE ROYAL SOCIETY

IT has been the fashion of late to decry the alleged influence of Bacon on the advance of science, but to the establishment of this fact we have the direct testimony of men of science living soon after his time, while he himself expressly left his " name and memory " to " the next age." Certainly there was on all sides a quickened spirit of inquiry and, in many different sciences, more students than ever were going direct to Nature as the source of our material knowledge. As men realised the extent of the unknown, the old attempt at universal knowledge was recognised as impossible. While the individual worker specialised, he gladly associated himself with those who were investigating other fields of learning and might well assist him in the search for truth.

Before the middle of the seventeenth century, this wish for co-operation in scientific work had shown itself in Naples, Rome, Paris, and London. Private meetings of men of science from 1630 developed in 1666 into the French *Académie des Sciences*. The corresponding movement in England, though beginning later, became organised somewhat earlier. Germany and Florence followed suit before the end of the century. Gresham's bequest of his house in Basinghall Street and an endowment for daily lectures by resident professors on the whole circle of the sciences might well have developed into a teaching university for London —it did, in fact, afford a nucleus for the Royal Society. In 1644, Robert Boyle returned from Geneva, a lanky, thoughtful youth of eighteen, and turning away by the advice of his sister, Lady Ranelagh—the friend and protector of Milton —from the strife of parties, made the acquaintance of

Hartlib. In the same year Milton dedicated to Hartlib his *Tractate on Education,* sketching the formation of an Academy, joint school and college. Three years later William Petty addressed to the same person a remarkable scheme for a technical college with a three-years' course for apprentices. In 1651 Hartlib himself followed, with a proposal for an agricultural college with a seven-years' course for youths from fifteen to twenty-two. In 1659 Evelyn put forward his scheme for a " philosophic mathematic college . . . somewhat after the manner of Carthusians," for adult philosophers, with its " elaboratory, physick garden, and conservatory." In 1661 Cowley published his proposal for a philosophical college with travelling professors combining research with teaching. All these suggestions equally exhibit the notion of combined work for the acquisition of knowledge that was then general.

In 1645 a little club of " worthy persons inquisitive into natural philosophy " began to meet weekly, either at the house of Dr. Jonathan Goddard in Wood Street, Cheapside, at the neighbouring Bull's Head Tavern, or at Gresham College. Among them were Dr. John Wilkins—afterwards brother-in-law of Cromwell and Bishop of Chester—Dr. John Wallis—clerk to the Westminster Assembly—and William Petty; to their meetings Hartlib soon brought Boyle. Their business, as one of them afterwards expressed it, was:

> " Precluding matters of theology and state-affairs, to discourse and consider of philosophical enquiries, some of which were then but new discoveries, and others not so generally known and embraced as they now are; with other things pertaining to what hath been called the New Philosophy, which from the times of Galileo at Florence, and Sir Francis Bacon (Lord Verulam) in England, hath been much cultivated."

In 1648 Wilkins became Warden of Wadham College, Oxford, and a year later Wallis became Savilian Professor of Geometry. Goddard afterwards became Warden of

Merton College; Petty was also at Oxford, where also was
"that prodigious young scholar, that miracle of a youth,"
Christopher Wren, then Fellow of All Souls. Boyle, a man
of large independent fortune, naturally established himself
at Oxford, though not until 1654. The weekly meetings,
begun in London, were resumed first at Wadham College
and then in Boyle's rooms. To Oxford, too, came one whom
Miss Mason, in her fascinating biographical sketch of Boyle,
describes as "a little deformed man, with a pale, sharp,
clever face, and lank dark hair that hung about his eyes;
a man with a stooping figure and a quick step; a queer little
solitary man who ate little and slept less, and worked rest-
lessly and incessantly; a man, even in those young days,
of a melancholy, jealous temper, warped by ill health. This
was Robert Hooke, who had come to Oxford in 1653, when
he was eighteen, as servitor or chorister at Christ-church."

This sickly youth, the son of a clergyman, was born at
Freshwater, in the Isle of Wight. Too delicate for regular
learning, he had used his clever fingers and brain in making
wonderful mechanical toys. He had been under Busby at
Westminster, and now, on the recommendation of Willis,
became personal assistant to Boyle. Hooke's versatility was
as remarkable as his industry. He constructed air-pumps,
diving-bells, hygroscopes, wind-gauges, and rain-gauges;
but perhaps his greatest service was to biology when he
made the microscope into a scientific instrument.

Roger Bacon in 1276 had explained the use of convex
lenses, whether as spectacles or as magnifying glasses. When
Galileo had heard of the invention of the telescope in Holland,
about 1608, he not only constructed one for himself, but also
at once recognised the possibility of its use as what we now
term a compound microscope. The first compound microscope is said to have been brought to England from Holland
in 1619 by Cornelius Drebbel, who made others in this
country.

Hooke made and used both simple and compound micro-
scopes, but it is remarkable how much excellent and detailed

work was done by botanists, almost down to our own time, with the simple lens. The little dissecting microscope with which Robert Brown did most of his work, not only in anatomy but also in histology, is still in use in the Natural History Museum. A similar instrument, which was Bentham's sole assistance in his systematic observations, is preserved in the Kew herbarium. Although Malpighi was, perhaps, the first person to use the microscope for animal anatomy and histology, Hooke was the first to apply it to the study of plants.

In 1657 Christopher Wren had become Professor at Gresham College and, after the Restoration, it was at the conclusion of one of his Wednesday afternoon lectures that a little gathering of those who had already for years been working together, resolved to start formally a society " for the promotion of Physico-Matthematicall-Experimentall Learning." There were present Boyle, Wilkins, Goddard, Petty, Wren, and Lord Brouncker, the first President. Thirty-three names were at once suggested, including Wallis, Willis, Evelyn, the poets Cowley and Denham, the Catholic Sir Kenelm Digby, and Henry Oldenburgh, who had been tutor to Boyle's nephews and who became the first secretary. Among the distinguished names soon after added to the list are those of Hooke, Dryden, Waller, Aubrey, Ashmole, Francis Willughby, and Isaac Barrow.

As John Richard Green puts it: " Science suddenly became the fashion of the day. Charles was himself a fair chemist. The Duke of Buckingham varied his freaks of rhyming, drinking, and fiddling by fits of devotion to his laboratory. The curious glass toys called Prince Rupert's drops recall the scientific inquiries which amused the old age of the great cavalry leader of the Civil War. Wits and fops crowded to the meetings of the new Society. Its definite establishment marks the opening of a great age of scientific discovery in England. Almost every year of the half-century which followed saw some step made to a wider and truer knowledge."

The Royal Society was granted a charter in July 1662, and Boyle allowed Hooke to leave his service to become its Curator of Experiments. In 1665 Oldenburg began the regular issue of its *Philosophical Transactions*, and in the same year Hooke published his *Micrographia*. This is a heterogeneous collection of microscopical observations from the edge of a razor and crystals of snow to the foot of a fly, the sting of a bee, the hygroscopic awn of the wild Oat, and the structure of wood and cork. Here it is that we have the first use of the word " cell " for the units of plant structure, in a comparison of cork with honeycomb. Moulds, leaf-mildew, and the stinging hairs of the Nettle are clearly represented in plates drawn by Hooke himself, and we have the happy suggestion that the movements of the leaves of the sensitive plant may be due to the circulation of internal liquid.

The youthful and enterprising Royal Society tried hard to secure the co-operation of anyone who might aid in the " advancement of natural knowledge." If its founders were mostly medical men, physicists and chemists, they were not blind to the importance of biology, and in 1667 they invited communications from Marcello Malpighi of Bologna.

Born near Bologna in 1628, Malpighi had been Professor of Medicine in the Universities of Bologna, Pisa, and Messina. He had demonstrated the existence and function of the capillaries connecting the arteries with the veins, the structure of the lung, the layer of the skin in which resides the colour of the negro—a tissue which still bears his name—and the surface papillæ of the tongue.

Malpighi responded to the suggestion of Oldenburg by a remarkable anatomical memoir on the silkworm, and he was elected a Fellow of the Royal Society in 1668. Most of his subsequent work was published by the Society. It is said that while walking in the garden of a friend, during 1663, his attention was directed to a broken stalk—probably a petiole—of the Chestnut from which thread-like strands— which we now term " vascular bundles "—projected.

Examining them with a lens, Malpighi discovered the vessels with a spiral thickening-layer, and this observation directed his attention to the study of vegetable histology. He published nothing on this subject until 1671, however, and this brought about one of those remarkable cases of simultaneous work by two independent observers; of which two of the best known examples are the discovery of the principle of natural selection by Darwin and Wallace, and the mathematical discovery of Neptune by J. C. Adams and Le Verrier.

Nehemiah Grew was born at Mancetter, Warwickshire, in 1641, his father, Obadiah Grew, being master of the Atherstone grammar-school, but becoming afterwards well known as minister of St. Michael's, Coventry, from which he was ejected by the Bartholomew Act of 1662. Grew went to Pembroke Hall, Cambridge, where he graduated B.A. in 1661. As Ray was then a don of Trinity College, it has been suggested that Grew may possibly have accompanied him on some of his " simpling " expeditions. Grew seems to have practised medicine at Coventry from 1664, although he did not take his degree of M.D. at Leyden until 1671. He himself tells us that in 1664—the year after Malpighi's observation on the Chestnut stalk—while studying animal anatomy, it occurred to him that as both plants and animal " came at first out of the same Hand, and were, therefore, the contrivances of the same Wisdom . . . it could not be a vain design to seek it in both."

He therefore watched the germination of seeds in his garden, noticing differences between that of Wheat and of the Bean, recognising the foliar nature of the cotyledons, and the existence in the seed-coat of the *foramen* which Tournefort afterwards named the micropyle, naming the descending axis the *radicle*, and the cellular tissues " extended much alike both in the length and breadth," *parenchyma*. He describes seed, root, trunk, leaves, flower, and fruit in order, and in the flower he distinguishes the calyx or empalement, the corolla or *foliation*, and the stamens or *attire*, each consisting of a filament or *chive* and an anther or *semet* containing a fine

o

powder of *globulets* (pollen-grains). None of these six last-mentioned terms ever passed into general use.

As to the uses of the pollen, Grew, in 1670, could only suggest " ornament or distinction to us, or food for other animals " (*i.e.* for insects). The account of these earliest researches was submitted—under the title of *The Anatomy of Vegetables Begun*—to Oldenburg, who showed it to Wilkins, then Bishop of Chester, and he laid it before the Society. In November 1671 it was sent to the printer and Grew was admitted a Fellow of the Society. This work, done for the most part without a microscope, appeared with three plates in the following month, and on the very same day Oldenburg received from Malpighi the manuscript of his *Anatomes Plantarum Idea*, which, however, was not accompanied by drawings. This introductory sketch formed fifteen folio pages in the volume in which it was afterwards incorporated, and deals with the histology of the stem, in which Malpighi describes bundles of fibres, tracheæ, spiral vessels, the horizontal structures now known as medullary rays and, in *Ficus*, the tubes containing the milky juice (laticiferous vessels). Unfortunately, misled by a preconceived analogy, he imagined a peristaltic action in the spiral vessels.

Encouraged at first by the reception of his work, Grew next wrote a detailed scheme of his whole design, but hearing that so distinguished an anatomist as Malpighi was engaged on the same task, he laid it aside.

Wilkins secured his appointment by the Royal Society as Curator of the Anatomy of Plants at a salary of fifty pounds ; and accordingly he established himself in 1672 as a physician in London and once more took up his researches. In 1673 he published his *Idea of a Phytological History Propounded, with an Account of the Vegetation of Roots*, with seven plates to this latter part. His aim in this work, unlike Malpighi's purely histological object, was a basis for a natural system of classification which was to include ecology, properties, chemical characters, and both external and internal anatomy. In the same year Grew writes to Malpighi acknowledging that he had first learnt of the existence of spirally-thickened tracheæ from the *Anatomes Idea*.

In 1673 and 1674 Grew laid his *Comparative Anatomy of Trunks*, with 18 plates, before the Society; and it was not until January 1675 that the first part of Malpighi's *Anatomes Plantarum* reached England. Both works were published during that year, Malpighi's being illustrated by 54 copper-plates engraved in London from exquisite drawings in red chalk, apparently by the author himself. Both authors had obtained a good general notion of the stem of the Maize with its scattered bundles and absence of distinct bark or pith, of the medullary rays in Dicotyledons, and of resin-passages and laticiferous vessels. Malpighi names the sap-wood *alburnum* and recognises the formation of annual rings of wood, although not understanding the nature of the cambium. He figures stomata for the first time, describing them as discharging either a vapour or a liquid into the air and, following Aristotle in comparing the fall of the leaf with the moulting of animals, he arrived by argument at a very fair notion of leaf-function and plant-nutrition gener-ally, though ignorant of the action of chlorophyll. He believed that water and dissolved substances, absorbed by the roots, ascended by the wood-fibres, air for respiration being conveyed by the spirally-thickened tracheæ; that the leaves elaborated the crude sap, giving off superfluous water, like the skin of an animal; and that the elaborated sap passed downwards by the laticiferous vessels to nourish new growths or to be stored in the cortex and pith-rays. He gives a good account of the development of the seed and embryo, with excellent figures showing both the embryo-sac and the endosperm. The second part of the *Anatomes Plantarum* was published in 1679 with 39 plates, and the Society collected and reprinted unaltered the author's numerous papers in two volumes in 1686 and 1687, and in a posthumous volume in 1697. In 1684 Malpighi's house and all his books and manuscripts were destroyed by fire; after this he seems to have done little original work. In

1691 he removed to Rome as physician to Pope Innocent XII, and there he died in 1694.

Meanwhile Grew had steadily carried on his independent work. *The Anatomy of Leaves, prosecuted with the bare Eye, and with the Microscope*, was laid before the Society in October 1676, and in it he speaks of the stomata (without mention of Malpighi's description published in the previous year) as " orifices or passports, either for the better avolation of superfluous sap, or the admission of air." Before the end of the year *The Anatomy of Flowers* was read; and in 1677, *The Anatomy of Fruits* and *The Anatomy of Seeds*. These three last-named portions of the work were not printed until 1682, when they appeared, with a revision of all his earlier octavo memoirs, in folio form as *The Anatomy of Plants*, with 83 plates. Thus it was in 1682 that the conversation between Grew and Sir Thomas Millington, which first enunciated the true function of the essential organs of the flower, was made public, but of this we shall speak in a later chapter.

In 1677, on the death of Oldenburg, Grew became Secretary to the Society, and his later works do not concern us here. He died in 1712. It is abundantly clear from their correspondence that the most cordial relations always existed between the two great founders of plant-histology. Misunderstanding as to dates of publication by various German authors led to charges of plagiarism against Grew. Unfortunately Sachs's *History of Botany*, translated with all its errors by Garnsey and Balfour in 1890, while endeavouring to do justice in this case, has fallen into similar mistakes. Many of the dates in that work are erroneous.

Like Malpighi, Grew fails to recognise the cambium, and both workers are constantly led astray by relying on supposed analogies with animal anatomy. Malpighi, for instance, while accurately figuring tyloses in the vessels of plants, compares them to the air-vessels in the lungs. Grew, amongst other valuable observations, notes the alteration of the floral whorls; describes a bulb as " a great bud underground "; and gives an account of winged and plumed fruits and seeds and of the explosive dispersal of seed. The beautiful and accurate draughtsmanship of the plates by these two seventeenth-century workers is a constant source of surprise to those seeing them for the first time, who may very likely have been previously under the delusion that plant-histology was "made in Germany" in the nineteenth century.

One supposedly early histological record of the Royal Society proves to be but an amusing mare's-nest. It is recorded in the history of the Society that "Mr. Henshaw exhibited the spirals of nut trees, showing that they grow snail-wise." This was rendered into German and then into French and back into English, in history of botany after history, from Sprengel to Sachs, as his discovery of spiral vessels in the Walnut, whereas to anyone looking at the original record it must be obvious that what were exhibited were some stems of the Hazel distorted by Honeysuckle!

CHAPTER XXVIII

JOHN RAY

SIDE by side with the foundation of plant histology by Hooke, Grew, and Malpighi, under the auspices of the Royal Society, work of equal importance was being accomplished—more or less in connection with the Society—in descriptive and systematic Botany.

John Ray (Plate IX), son of a blacksmith, was born at Black Notley, Essex, in 1627. He obtained his first education at the grammar school (which at that time was held in Braintree church), and in 1644 was sent to Catherine Hall, Cambridge, at the expense of a neighbouring squire. In 1646 he migrated to Trinity College and in the year following he graduated; and in 1649 he was elected Fellow of his college. Two years later he proceeded to his M.A. degree and was made Greek lecturer. In 1653 he became mathematical lecturer and in 1655 humanity (*i.e.* Latin) reader. It is stated that an illness led to his taking up the outdoor study of Nature as a healthy relaxation. This illness was possibly the beginning of the tuberculosis from which he undoubtedly suffered in later life, and was aggravated, perhaps, by over-work.

Ray's *Wisdom of God manifested in the Works of the Creation* and *Physico-Theological Discourses concerning the Dissolution and Changes of the World*—which anticipate the teleological theism of Paley—though not published until 1691 and 1692, were preached as college exercises before his ordination. It was during the long vacation of 1658 that he made the first of his mainly botanical tours—a record of which we have in his journals—travelling on horseback through the Midland counties and North Wales.

JOHN RAY (1627-1705).

Plate VIII.

facing p. 204.

In 1660 he published his first work, a modest little duodecimo of 285 pages containing an alphabetical catalogue
of the plants of Cambridge, with synonymy, notes on their
uses, and a glossary. This was the first regular English
florula, if we do not reckon Johnson's *Ericetum Hamstedianum*
as being one, and is a model for its scrupulously painstaking
accuracy. " I resolve," Ray writes a few years later,
" never to put out anything which is not as perfect as is
possible for me to make it."

He was ordained in the same year, but remained at Cambridge as a resident fellow and tutor, making two further
summer tours in England, Scotland, and Wales. On the
third of these, in 1662, he was accompanied by his pupil,
Francis Willughby, eight years his junior, who was for the
next ten years to be his intimate fellow-worker. Then
came the Bartholomew Act, and Ray—who had never
taken the Solemn League and Covenant, which he even
considered an unlawful oath—could not conscientiously
declare it to be not binding on those who had taken it.
He accordingly resigned his fellowship and retired into lay
communion with the Established Church, in which he was
never thereafter able to seek preferment.

Agreeing to divide between them the task of describing
the whole known organic world, Ray undertaking the plants
and Willughby the animals, the two friends started in the
spring of 1663 on a continental journey that occupied three
years. On their return Ray became domiciled in Willughby's
home, Middleton Hall, in Warwickshire, and here he
engaged himself in classifying Willughby's collections. He
also prepared with him the synoptic tables of animals and
plants for which Wilkins, whom we have previously spoken
of as the patron of Grew, had asked in his *Essay towards a
Real Character and a Philosophical Language.*

Wilkins, whom his wife's nephew, Richard Cromwell, had
made Master of Trinity, had then regained Court favour and
was made Bishop of Chester in 1668, the year in which this
curious *Essay* was published. His proposal was for the

formation of an international scientific nomenclature, and the tables that Ray prepared are of interest as the first directly systematic work published in England and as the foundation of his *Methodus* of 1682. He begins with the unfortunately traditional distinction into Herbs, Shrubs, and Trees. The two latter groups are subdivided mainly by their fruits, cone-bearing, nut-bearing, pomiferous, pruniferous, and bacciferous trees and " siliquous " (*i.e.* leguminous) shrubs being recognised as subdivisions, as also are spinous and " sempervirent " shrubs. Herbs are divided into groups in which leaves, flowers, and seed-vessels afford the best group-characters. In this manner the cereal and other Grasses, the bulbous *Monocotyledons*, *Boraginaceæ*, *Stellatæ*, *Polygonaceæ*, and other *Incompletæ*, and *Cruciferæ*, are kept fairly well together. This tentative system was slightingly and not very fairly criticised by Morison, as if based entirely on leaf-characters. This criticism was the beginning of unfriendly relations between the two botanists, members of rival universities and of opposite political parties.

Robert Morison (Plate X) was born at Aberdeen in 1620 and is the first of the long line of distinguished Scottish botanists. He graduated at Aberdeen in 1638, but—as in the case of Thomas Johnson—on the outbreak of the Civil War he espoused the Royalist cause and was wounded in 1644 at the Brig of Dee. With other Royalists he fled to France, where he studied the medical sciences at Angers and graduated as M.D. in 1648. Vespasian Robin, Botanist to the King of France—whose name is commemorated by the *Robinia*, which he introduced—recommended him in 1650 to Gaston, Duke of Orleans, and he became one of his physicians and garden-curators at Blois. Here he may have taken part in the preparation of the first two editions of the *Hortus Blæsensis* (1653 and 1655); while the third, published in 1669 after the death of Brunyer, seems to have been wholly his work. The Duke introduced Morison to his nephew Charles II, who at his Restora

ROBERTUS MORISON M.D
Natus Aberdoniæ   Obiit Londini
Anno 1620.   Anno 1683.

Quæ Morisone viro potuit contingere major (Ipse tibi palmam Phæbus concedit Apollo,
Gloria, Pæonium quam superasse genus) Lauroaque est capiti quælibet herba tuo.
                                          Archibaldi Pitcairne M.D.

ROBERT MORISON (1620-1683).

facing p. 206.

tion summoned him to England, employed him in laying
out St. James's Park and gave him the titles of Royal
Physician and Botanist, with a nominal salary of two
hundred pounds a year. In 1669 he was made Professor
of Botany at Oxford, where during term time he was in the
habit of giving three demonstrations weekly, at a table in
the middle of the newly-established Physick Garden. In
1683 he was knocked down by the pole of a coach while
crossing from Northumberland House to St. Martin's Lane,
died at his house in Leicester Fields, and was buried in
St. Martin's Church.

Morison's first work, *Præludia Botanica* (1669), is practically
three distinct treatises, though with one pagination, forming
altogether a small octavo of about 500 pages. These three
are the catalogue of the Blois garden, a very severe criticism
on the Bauhins, and a dialogue on classification. In a
dedication to King Charles, Morison states that he had
drawn up " a system, new and derived from Nature, which
(without boasting) has been observed by (him) alone,
although coeval with the beginning of the world," and that
the Duke of Orleans had promised to help him to publish
it. The *Dialogue*, professedly between a Fellow of the
Royal College known as Gresham's and the King's Botanist,
lays down the dictum that group characters should be taken
from flower and fruit, not from properties or from the form
of leaves, and then trounces the system put forward by Ray
as mere chaos.

Morison then planned a great systematic work to be
styled *Plantarum Historia Universalis Oxoniensis*, and in 1672
published as a specimen a thirtieth part of the entire scheme,
Plantarum Umbelliferarum Distributio, a folio of ninety-one
pages with 150 figures on twelve plates. This excellent
piece of work was the first monograph of a group of plants
ever published, if we except Belon's *De arboribus coniferis* of
1553. This Morison followed by the publication (in 1680)
of the second part of the *Historia*, 617 pages, dealing with the
first five sections of herbaceous plants, viz. Climbers,

Leguminosæ, *Siliquosæ*, " *Tricapsulares Hexapetalæ*," i.e. the
Petaloid Monocotyledons, and another group of "Tricapsulares." After his death, Jacob Bobart the younger,
Keeper of the Physick Garden, having learned the system
from its author, completed the part relating to Herbs in a
volume of 655 pages with figures of nearly 3400 plants
engraved on copper, and published in 1699. Not even then
had any complete view of Morison's system appeared, however, nor did it until after the death of both Bobart and
Ray, when, in 1720, a twelve-page tract, probably by
Bobart, was printed at Oxford under the title of *Historiæ
Naturalis Sciagraphia*. In this synopsis, now exceedingly
rare, we have the fourfold main division into trees, shrubs,
under-shrubs, and herbs. The trees and shrubs are grouped
by their fruits, very much as in Ray's Tables of 1668; the
under-shrubs are all climbing plants; and, among the
fifteen herbaceous sections, besides the groups above-
mentioned as described in 1680, the *Campanulaceæ*, *Scrophu-
lariaceæ*, *Rutaceæ*, *Geraniaceæ*, *Malvaceæ*, *Caryophyllaceæ*, *Primu-
laceæ*, *Polygonaceæ*, *Chenopodiaceæ*, *Compositæ*, Grasses, Sedges,
Umbelliferæ, *Stellatæ*, *Euphorbiaceæ*, *Labiatæ*, *Boraginaceæ*, *Orchid-
aceæ*, *Filicineæ*, and Cellular Cryptogams are kept together
in this order.

Morison had by no means kept to his own principle of relying on the flower and fruit alone. Tournefort in 1694 writes:
" We do not know how to praise this author sufficiently;
but it seems that he praises himself a little too much. . . .
He dares to compare his discoveries to those of Christopher
Columbus, and without mentioning Gesner, Cæsalpinus or
Columna, asserts repeatedly that he has taken his matter
from Nature only. This one might believe if he had not
transcribed entire pages from the last two authors."

Linnæus says roundly: " All that is good in Morison is
taken from Cæsalpinus." As we have already mentioned,
a copy of that author, annotated by him, is still to be seen
at Oxford.

Ray was persuaded to join the Royal Society in 1667, and

he and Willughby took part in a series of experiments as to
the flow of sap in trees. The experiments were, perhaps,
less satisfactory than those of Hales in the next generation,
because of the excessive influence of the analogy of animal
circulation.

Willughby's marriage made no difference in the friendly
collaboration of the two naturalists, save that Ray made
one or two summer tours accompanied only by Thomas
Willisel, an old Cromwellian soldier, who acted as collector
to several of the botanists of the time. In 1670 Ray brought
out a *Collection of Proverbs* and his *Catalogue Plantarum Angliæ*,
which, although it only contains 1050 species in alphabetical
order, is far more trustworthy than longer lists by others.

The premature death of Willughby (in 1672) left to Ray
the completion of all the works upon which they had been
jointly engaged, and the education of his friend's three little
children, whilst an annuity of £60 was to constitute his main
income for the remainder of his life. In 1673 he married a
young woman twenty years his junior, who seems to have
been a nursery governess at Middleton, and published an
account of his foreign tour with a list of the plants observed.
It is somewhat amusing to find him in 1675 preparing a
Nomenclator Classicus, or English, Latin, and Greek vocabu-
lary, for his pupils then aged five and four. It proved so
useful, however, as to go through five editions by 1706.

By 1676 he had completed Willughby's *Ornithologia*—
which has been authoritatively styled the first serious attempt
at the classification of birds since the days of Aristotle and
" the foundation of scientific ornithology "—and it is impossible to separate in it the work of Willughby from that
of Ray. On the death of their grandmother (in 1676)
Ray's pupils were taken from him, and for the next two or
three years he was apparently much distracted by several
moves. Three years later, however, he settled at his native
village, in a house he had built for his mother, and this was
his home for the remaining twenty-five years of his life.

Not until 1682 have we any other work from Ray's pen;

but the *Methodus Plantarum* of that year, an elaboration of the tables prepared for Wilkins fourteen years before, is one of the corner-stones of his fame. In it he describes the true nature of buds, speaking of them as annual plants springing from the old stock. He recognises, though without naming them, the fundamental division into Dicotyledons and Monocotyledons. He has only three main groups: herbs in 47 " summa genera "; trees in 8, and shrubs in 6. Although the system is based mainly upon the fruit, it makes use also of characters derived from flowers and leaves. He freely acknowledges his indebtedness to Cæsalpinus—whom he styles " the parent of system "—to Jung and to Morison. " His system," says Lindley, " when altered and amended, as it subsequently was by himself, unquestionably formed the basis of that method which under the name of the system of Jussieu is universally received at the present day."

The death of Morison in 1683 turned Ray's attention once more to his share of the scheme he and Willughby had planned and which he had laid aside from unwillingness to compete with Morison's undertaking, viz. a general history of plants. Accordingly, having published (in 1686) Willughby's *Historia Piscium*, a folio of 370 pages, more than half of which was his own work, in the same year he produced the first volume of his *Historia Plantarum*, containing nearly 1000 pages. The second volume, of equal bulk, appeared in 1688, and the third in 1704. For each of these volumes Ray received from the publishers £30, or about threepence per folio page! The three volumes contain descriptions of over 18,000 species, as against 3500 in Jean Bauhin's *Historia*, thirty-five years before, and about 10,000 known to Tournefort, his own contemporary. Of far greater importance than this wealth of descriptive matter, however, is the introduction " *De Plantis in Genere*." It occupies the first 58 pages and is a masterly presentation of all then known of vegetable histology, anatomy, and physiology. Though confessedly based upon the work of Grew, Malpighi, Cæsalpinus, and Jung, it is so judicious as to be practically both original and authoritative. Linnæus studied it with care and Cuvier expressed the wish that it might be published separately. With characteristic caution Ray here only accepts Grew's conclusion as to the sexuality of plants as being " very likely." Grew, as we have seen, had been quite in doubt as to the use of pollen in 1671, although in his *Anatomy of the Flower* he writes (in 1676) :

> " In discourse hereof with our learned Savilian Professor, Sir Thomas Millington, he told me he conceived that the attire doth serve as the male for the generation of the seed. I immediately replied that I was of the same opinion: and gave him some reasons for it."

As Millington is otherwise unknown as an original investigator, we may follow Ray in attributing the first suggestion to Grew.

In 1690 appeared the *Synopsis Methodica Stirpium Britannicarum*, the first systematic Britisʰ flora, which, with the *Methodus* and the *Historia*, may be said to form the triple tiara of Ray's botanical fame. By that date he was fully convinced that the stamens were male organs. This book, with its two later editions of 1696 and 1724, was for more than seventy years the pocket companion of every field botanist in Britain. In 1693 Ray followed it with a *Synopsis Quadrupedum et Serpentini generis*, and in 1694 by a *Stirpium Europeanarum Sylloge*, his *Methodus Insectorum*, *Historia Insectorum*, and *Synopsis Avium et Piscium* appearing after his death. He also contributed most of the lists of plants in the counties of England that appeared in Gibson's edition of Camden's *Britannia* between 1693 and 1695.

To the second edition of his *Synopsis Stirpium*, published in 1696, Ray added a letter to Rivinus (1652–1723) on classification, in which he condemns that botanist's reliance upon the corolla as a sole character. At the same time he issued a tract on the principles of classification and the method employed by Tournefort (1656–1708) in his *Elemens de Botanique* (1694). These were preparatory to his revision of his *Methodus*. This, though finished by 1698, he could not get published until 1703. Then, although it was printed at Leyden, the publishers fraudulently put " London " on the title-page. In this *Methodus Emendata et Aucta*, Ray combined the Shrubs with the Trees, divided Herbs into *Dicotyledones* and *Monocotyledones*, and practically makes the same division among Trees. He now fully recognised that to attempt a grouping by any single set of characters inevitably broke up natural groups. He therefore took whatever characters best served in various groups, whether derived from flower, fruit, or seed. It is thus quite inadequate of Linnæus to say, as he does, that Ray merely from a " fructist " became a " corollist."

For many years Ray had suffered from painful ulcers in the legs, aggravated perhaps by insufficient food. In the *Methodus emendata* he expresses his regret at his inability to visit London botanic gardens and herbaria. He speaks of himself as a thin body, subject to colds and one whose lungs were apt to be affected while his cottage was exposed to north-east winds. As early as 1693 he speaks of sleeplessness, and in November 1704 he writes to his friend Sloane that he doubted whether he would live through the winter. Among the Sloane MSS. in the British Museum is a touching farewell letter addressed to Sloane on January 7, 1705, which breaks off abruptly. Ten days later this great naturalist, of whom it has been well said that he became, " without the patronage of an Alexander, the Aristotle of England and the Linnæus of his age," passed away.

CHAPTER XXIX

MANY able naturalists visited the humble cottage of John Ray at Black Notley, or maintained a correspondence with the great naturalist who dwelt there. Among his visitors was his neighbour, pupil, and medical attendant, Samuel Dale, afterwards author of the *Pharmacologia*, who rendered him material assistance in critical British botany. The pugnacious Henry Compton, Bishop of London, who had escorted the Princess Anne on her flight from her father's Court and had filled the gardens of Fulham Palace with rare exotics, also made a pilgrimage to this botanical shrine. Thither too came Petiver, a London apothecary whom Ray terms " *mei amicissimus*."

Born and educated at Rugby, Petiver was in business in Aldersgate Street and took an active part in the management of the Physic Garden of the Society of Apothecaries at Chelsea. He made an extensive miscellaneous collection of natural history objects, and circulated printed lists of his desiderata and directions for collecting, among ships' captains and surgeons. In this way he formed a museum of stuffed animals, birds, insects, seeds, and plants, for which Sir Hans Sloane offered him four thousand pounds. Elected F.R.S. in 1695, Petiver began a series of illustrations to Ray's *Synopsis*, and proposed to do so also for the *Historia*, and for his various publications he prepared upwards of 300 copper plates with over 10,000 figures. Petiver acted as apothecary to Sloane, and on his death (in 1718) his collections and correspondence were incorporated with those of the great physician.

Among Ray's correspondents were Samuel Doody, also a

London apothecary, " the top of all the moss-croppers,"
Martin Lister, the zoologist, Sir Tancred Robinson, physician
and Latinist, and Sloane.

At first sight there does not seem to be any obvious connection between the bark of trees collected by Indians on
the Andes and the existence of a botanical garden at Chelsea,
and of the British Museum at Bloomsbury. In 1638 the
Countess of Chinchon, wife of the Governor of Peru, was
cured of an intermittent fever by an infusion of the bark
of a Rubiaceous tree of those regions, of a genus which has
since been known as *Cinchona*. The use of this drug in a
crude form was disseminated throughout Europe by the
Society of Jesus, and thence it came to bear the alternative
names of " Peruvian " or " Jesuits' Bark." By far the
greater number of ailments in the undrained England of
the seventeenth century were of a rheumatoid or aguish
nature, and thence it came about that physicians reaped
large fortunes by the prescription of this new and efficacious
tonic bitter. Terrible, too, were then the ravages of small-
pox, and inoculation, the crude approximation to our
modern vaccine treatment, did much to check them.
Sloane's fame and fortune as a physician depended largely
upon " bark " and inoculation, and to that fortune in the
main we owe the Chelsea Physic Garden and the British
Museum.

Sloane (Plate XI), born in 1660 at Killileagh, co. Down,
was of Scottish race. Showing signs of lung trouble when
sixteen he became an abstainer, to which fact he attributed
his long life. He studied medicine for four years in London,
during which time he made the acquaintance of Boyle and
Ray, and came to know the Chelsea Garden, which he
afterwards so greatly benefited. He then went to Paris
with Tancred Robinson, and here he studied under Tournefort. He next moved to Montpellier, where Magnol,
teacher alike of Tournefort and of Bernard de Jussieu, was
Professor of Botany. Sloane graduated as M.D. at Orange
in 1683, and on his return to London became a resident

HANS SLOANE (1660-1753)

facing p. 214.

Plate X.

assistant to the celebrated Sydenham; in 1685 he was elected a Fellow of the Royal Society.

In 1687 Sloane accompanied the Duke of Albemarle to the West Indies as physician, but the death of the Duke restricted his stay in Jamaica to fifteen months, and in 1687 he was back in London with some 800 species of West Indian plants. He set up in practice in Bloomsbury Square and soon became popular. From 1693 to 1712 he acted as Secretary to the Royal Society, and in 1695 he married the daughter of Alderman Langley. The publication, in 1696, of his *Catalogus Plantarum in Insula Jamaica*, classified according to Ray's system, gave him considerable scientific reputation. He furnished Ray with a list of English plants introduced into Jamaica, for the *Synopsis* of 1696, and with the use of all his manuscripts as to Jamaica for the third volume of the *Historia*. His own larger work on his West Indian *Voyage* appeared in two folio volumes in 1707 and 1725, and is dedicated to Queen Anne, who was one of his patients. It was Sloane who suggested bleeding in the Queen's last illness, and he also inoculated several members of the royal family for small-pox. In 1712 he purchased from Lord Cheyne the manor of Chelsea, then a suburban village, but it was not till 1736 that he took up his residence in the house that had once been the home of Sir Thomas More. Being a Whig, he was made Physician-general by George I in 1714 and a baronet—the first medical man to be so honoured—in 1716.

From the death of Newton in 1727 until 1741 Sloane was President of the Royal Society, and in 1742 his great collection of books, manuscripts, dried plants, and other natural history objects and antiquities, was removed to Chelsea Manor-house. He died at Chelsea in 1753, bequeathing his collections to the nation for £20,000, less than half what they had cost him. As he had for years bought up the herbaria of deceased botanists, his herbarium, which consists of 310 large volumes, is now a priceless key to pre-Linnæan botany, while his collection of manuscripts is a

P

storehouse of even greater general interest. The Government accepted the bequest and purchased Montague House, Bloomsbury, for its reception, thus forming the nucleus of the British Museum.

We have already recounted something of the great debt that English botany owes to individual members of that Society, from Thomas Johnson onwards. The Society was incorporated by Royal Charters in 1606 and 1617, and at an early date they began those "herborisings" which, until their abolition in 1835, did so much to encourage botany among London medical students. In 1673 they took a lease from Charles Cheyne of about four acres of ground at Chelsea, partly for the erection of a barge-house. A "stove" was, however, erected in 1681 and, in 1683 the four cedars were planted near the river-front—trees which since 1904 have only been represented by five chairs at the Society's Hall. Evelyn records a visit to this "Garden of Simples" in 1685, and Doody and Petiver had acted as managers and demonstrators of plants before Sloane's gift of the freehold.

At that time Philip Miller, "*hortulanorum princeps*," was appointed gardener, and he retained that post until 1770, a year before his death. Miller was of Scottish origin and was born in 1691, fourteen years before the death of Ray, and had probably been employed in the Chelsea Garden in his boyhood. Writing in 1790, Pulteney says he was "the only person I ever knew who remembered to have seen Mr. Ray. I shall not easily forget the pleasure that enlightened his countenance, it so strongly expressed the *Virgilium tantum vidi*, when, in speaking of that revered man, he related to me that incident of his youth."

The incident in question was probably one of Ray's rare visits to the Chelsea Garden during the later years of his life.

In his "apartments in the greenhouse" at Chelsea, Miller compiled his excellent *Gardeners' Dictionary*, of which Linnæus said, "*Non erit Lexicon Hortulanorum, sed Botanicorum*," the

first edition appearing in two octavo volumes in 1724, the
second in one folio volume in 1731. Brought up to use
Ray's system, it was not until the seventh edition, in 1759,
that Miller could be persuaded to adopt that of Linnæus.
After his death, Professor Thomas Martyn, who had been
born in 1736, close by in Church Street, brought out the
ninth edition " in four huge volumes, ' with plates,' " which
the old squire's scotch gardener in *Mary's Meadow* pronounced to be " a gran' wurrk." Here too in 1736, Linnæus, then twenty-nine, visited Miller, who, he says " permitted me to collect many plants in the Garden, and gave
me several dried specimens collected in South America."
Here again, many years later, came as a student the wealthy
young Lincolnshire squire Joseph Banks.

Since 1899 the Garden has ceased to belong to the Society
of Apothecaries; but it retains much of its old-world charm.
If we find our way down Swan Walk—which once led to
the " Old White Swan," a riverside hostelry favoured by
Pepys and by Dibdin—we come to the simple iron gateway,
through which we see the trim lawns and orderly beds of
the old garden, whose successive rearrangements have
reflected the history of systematic botany from Ray to
Linnæus and from Linnæus to Lindley. As the presiding
genius loci in its centre still stands the marble statue of
Sloane by Rysbrach, which the Society of Apothecaries
erected, " With grateful Hearts and general Consent," in
1737.

A house nearly opposite the gateway, No. 4, Swan Walk,
is associated with a pathetic romance of a brave woman
and an unfortunate adventurer; for there, in 1735, came
Elizabeth, wife of Alexander Blackwell. The husband,
born in 1709, was second son of the Principal of Marischal
College, Aberdeen. Though he acquired a good knowledge of Latin and Greek at an early age, he seems to have
left the University of Aberdeen without a degree, to have
eloped with the daughter of a well-to-do stocking-merchant
of Aberdeen, and to have started life in London as a corrector

for the press. About 1730 he set up as a printer, but was ruined by a trade combination against him because he had never been apprenticed to the trade. He spent two years in a debtor's prison, and it was then that his wife Elizabeth carried out her great undertaking. Having considerable skill as a draughtswoman and colourist, and hearing that an illustrated herbal was wanted, she, by the advice of Rand, then Demonstrator at Chelsea, took up her residence at 4, Swan Walk, that she might draw fresh plants from the Garden. Not only did she draw them, but she also engraved them on copper and coloured the prints with her own hands. With the author's consent, Blackwell in his prison supplied nomenclature and abridged descriptions from Joseph Miller's *Botanicum officinale* (1722), and in two years' time the first folio volume of *A Curious Herbal* was completed, with 250 plates, the second being published in 1739.

The utility of this work is, perhaps, best evidenced by the issue of a second edition, with 100 more plates, under the auspices of C. J. Trew of Nuremberg in 1750–73; but what was of more importance to the plucky authoress was that she released her husband. He then figured as an agricultural expert and published *A New Method of improving Clayey Grounds*, in consequence of which he was invited in 1742 to Sweden and given the charge of a royal model farm at Allestad. Linnæus, who visited him there, speaks of him in a work on *Divine Vengeance*, written for his son's guidance, as " a bold, ignorant atheist," brought over by Linnæus's friend Alströmer, after whom the genus *Alstræmeria* was named. No crime seems too great to be laid to Blackwell's charge. He is alleged to have poisoned a man whose wife he had seduced, to have suggested the murder of Alströmer and the minister Tessin to secure for England the ruin of Swedish trade, and finally to have conveyed to the weak king Frederic of Hesse a message from Louisa, Queen of Denmark, daughter of George II, that he should be provided with means to make himself absolute, if he would consent to alter the Swedish succession in favour of her brother the Duke of Cumberland. Whether guilty or merely a schemer, Blackwell was arrested, cruelly tortured by the pill and the rack, and condemned for high treason to be broken alive upon the wheel. This was commuted to decapitation, and he was beheaded at Stockholm in 1747. It is related that having laid his head on the wrong side of the block he apologised to the executioner, saying that it was the first time he had ever been beheaded! He is said to have sent remittances to the wife and child he had left in England, who were—it is also said—about to join him in Sweden; but nothing is known as to their subsequent history.

CHAPTER XXX

THE SEXES OF PLANTS AND SOME JACOBITE BOTANISTS

THE recognition by Grew and Ray of the universality of sex among Flowering Plants was specially significant in that it was distinctly contrary to the long-received Aristotelian theory. Although Ray had quoted (in 1686) the old example of the Date-palm and suggested that when the anthers and the stigmas are on distinct plants wind may carry the pollen, it was unquestionably Camerarius who first clearly demonstrated the matter by experiment.

Rudolph Jacob Camerarius was born at Tübingen in 1665 and graduated there in philosophy and medicine. After two years of European travel, during which he visited England, he became Professor and Director of the Garden in his native town, where he died in 1721.

In 1691 Camerarius noticed that, although no staminate tree was near, a female tree of the Mulberry (*Morus nigra*) bore fruit, and that the berries contained empty " seeds " (*i.e.* capsules). He then experimented by separating plants of the two sexes in *Mercurialis annua* with a like result, which he published in the journal of the Leopold-Caroline Academy for that year. He detailed, with much careful reasoning, his further experiments in a *Latin Epistola de sexu plantarum* (63 pp. 12ᵐᵒ) in 1694. He had removed the undeveloped anthers in *Ricinus* and cut off the stigmas in Maize. Influenced by the observations of Ray and Swammerdam upon the existence of androgyny in snails and other animals, however, he came to the very natural but erroneous conclusion that perfect flowers are self-fertilising, the state of things that is the exception among animals being the rule among plants.

In 1703 Samuel Morland, in a paper communicated to the Royal Society, proposed to extend to plants the view of sex propounded by the microscopist Leeuwenhoek in the case of animals. He suggested that the pollen is " a congeries of seminal plants," one of which must be " conveyed into every ovum before it can become prolific "; that they travel down the canal of the style; and that the unimpregnated seeds (ovules) have a perforation (the micropyle) by which the seminal plant enters.

The theory of the matter was but little advanced by Richard Bradley, Professor of Botany at Cambridge from 1724 to 1732. In his *Improvements of Gardening*, published in 1717, he alludes to the sterilising of Tulips by removing the anthers, and of Hazels by removing the catkins, to the production of hybrids among Auriculas, and between the Carnation and the Sweet William by artificial pollination. He recognised by microscopical examination that different forms of pollen-grain characterise different natural groups of plants, and suggested that the pollen being waxy was magnetically attracted to the stigma.

In popularising the new views, two men, who may not have added much to our knowledge, undoubtedly did much, the one in France and the other in England. These were Sebastien Vaillant and Patrick Blair.

Vaillant (1669–1722), curiously enough, had been a pupil of Tournefort, a strenuous opponent of the doctrine of sexuality in plants. He had just been appointed Professor of Botany and Director of the Jardin du Roi when that garden was reopened, after reorganisation, in 1717. On that occasion he delivered a *Discours sur la Structure des Fleurs* or *Sermo de Structura Florum*, which was published at Leyden, in French and Latin, in the following year. This address is noticeable for the first use of the term " ovary " with reference to plants. In language Pulteney condemns as " too florid " he describes his own observations on pollination, pointing out that the style is not an open tube, and that the pollen-grains certainly do not enter the micropyle, so that fertilisation is probably effected by a *souffle*, the " subtle and vivifick effluvium " of Grew, or the *aura seminalis* of Swammerdam.

William Sherard was the friend of Ray and Petiver, and afterwards founded the Sherardian professorship at Oxford. Like Vaillant he had studied under Tournefort, and in his honour Vaillant had named the genus *Sherardia*. In 1721 Sherard found Vaillant in feeble health and anxious about the publication of his remaining memoirs. Accordingly Sherard, with the assistance of their mutual friend, the illustrious Leyden physician Boerhaave, arranged for publication the folio *Botanicon Parisiense*, and this was issued in 1727 at Leyden and Amsterdam. Among the subscribers appear the names of Dr. Dale of Braintree, Haller, Dillenius, Mead, Rand, and Joseph and Philip Miller. As we shall see presently, the work contains interesting evidence of the extent and method of Vaillant's teaching as to sex.

Meanwhile, in 1720, Patrick Blair's *Botanick Essays* had appeared in London. Patrick Blair was born before 1670, probably in Dundee. Of his training we only know that he was in Flanders in 1695 and 1697, and was settled in medical practice at Dundee by 1701. In 1706 he gained considerable distinction by the dissection of an elephant, which he described for the Royal Society, and this led to the establishment of a local Natural History Society and a Physic Garden of which Blair was overseer. In 1712 he was elected F.R.S., and in 1713 he journeyed to London, visiting Sloane and Petiver's collections and returning by way of Oxford. Here Bobart had recently completed Morison's *Historia*, and here Blair inspected Tradescant's collections in the misnamed Ashmolean Museum.

The next we hear of him is his surrender as " Chirurgeon " with Lord Nairn's battalion at Preston on November 13, 1715, and his return to London between two Hanoverian troopers with hands and arms pinioned. He was lodged in Newgate, and there he was visited by Sloane and Petiver. He was put on his trial on March 31, 1716, and although

he insisted that he had been forced to act as physician and
surgeon to the Jacobite forces, he pleaded guilty and was
sentenced to death. Sloane's influence at Court was used
to secure his pardon, but, as the date fixed for his execution
drew near, the Doctor writes bitterly to Petiver of the
delayed reprieve. The even of his execution came; Petiver
and some other friends were with him; and, in a letter to
Sloané, Petiver tells how the " Doctor sat pretty quietly
till the clock struck nine; and then he got up and walked
about the room; at ten he quickened his pace; and at
twelve, no reprieve coming, he cried out ' By my troth, this
is carrying the jest too far.' " Soon after the reprieve and
pardon came, however, and the Doctor was spared for twelve
years more work.

He endeavoured to practise in London, and read before
the Royal Society a paper on the sexes of plants which he
was recommended to amplify into a volume. Then too he
made the acquaintance of John Martyn, a young clerk in a
mercantile house, who for nearly thirty years was to be
Professor of Botany at Cambridge and the father of Thomas
Martyn, editor of Miller's *Dictionary*, who held the same
Chair for sixty-four years more. In 1720 Blair moved to
Boston, Lincolnshire, where he died in 1728. During those
years Martyn revised his proofs for press—a duty that,
considering Blair's writing and spelling, can have been no
sinecure.

The *Botanick Essays*, which form an octavo volume of
414 pages, with four copper-plates, are five in number.
They deal respectively with the structure of the flower;
the fruit; a criticism of the various methods of classification; the generation of plants; and their nutrition. The
fourth essay occupies more than a quarter of the volume—
more than twice the length of Camerarius's *Epistola*. It
quotes Grew, Ray, Camerarius, Vaillant, and Bradley, and
describes and figures several plants upon which the writer
had himself experimented. It argues, as Vaillant had done,
that the style is not open, that the pollen does not enter the

micropyle, and that, therefore, fertilisation is probably effected by an effluvium. " For the Reader's Diversion," the essay concludes, " I have hereto subjoin'd an Ode written in Latin in Camerarius's Epistle *de Sexu Plantarum*, and literally translated by a young Botanick Student, which as it contains an Abstract of this Essay I have been advis'd to insert." And then follows:—" An Ode formerly Dedicated to Camerarius in Latin, and now presented to the Author. Being translated into English by J. Martyn."

In the face of this, Sachs in his *History of Botany* [1] has the audacity to write:—" Even the Latin ode is borrowed without acknowledgment " !

The language of the unknown Latin poet and of Martyn's English version may, perhaps, deserve Pulteney's censure of Vaillant's *Discours* as " too florid." The subject seems naturally to lend itself to poetical treatment. In Vaillant's *Botanicon Parisiense* (1727) there is an interesting Latin poem of 526 hexameters entitled *Fratris ad Fratrem de Connubiis Florum Epistola Prima*, and signed " Mac-encroe Hibernus, Medicinæ Doctor." On the two following pages, among Latin epigrams accompanying the portrait of Vaillant, are two signed " Demetrius de la Croix, Doctor Medicus." In the following year the same poem appeared separately at Paris in a 40-page pamphlet entitled " *Connubia Florum Latino Carmine Demonstrata Auctore D. De La Croix, M.D. Cum Interpretatione Gallicâ D. . . .*," which has as a frontispiece a rough illustration of the Passion-flower and the Scythian Lamb. In 1749 it was reprinted in the first volume of *Pœmata Didascalia*, where it is attributed to Patrick Trante, M.D. Apparently De La Croix is but the French version of the Irish name MacEncroe, and, as the British Museum library authorities conclude, Patrick Trante is merely the author of the French prose paraphrase in the Paris edition.

The Arch-Jacobite Atterbury, writing from Surennes in September 1728, sends his son-in-law eight copies of the poem, which, he says, was " writ by an Irishman here at

---

[1] English translation (1890), p. 391.

Paris, which, in some Parts of it, is excellent, and approaches very near to the manner of the Versification of Virgil's Georgics."

Two of the copies sent were "for Dean Swift and Mr. Pope."

As we might expect, the poem reflects the teaching of Vaillant. When the stamens burst, an *aura* or *souffle* passes by the canal of the style to "*les cordons ombilicaux*"; and "if the lovers are separated," *i.e.* in diœcious plants, the pollen is entrusted to the Zephyrs. London-made lenses are referred to, and *Ophrys, Papaver, Phœnix, Terebinthus, Parietaria, Stratiotes*, and *Mimosa* are among the chief plants employed as examples.

The poet describes how his illustrious master (Vaillant) took him to the most famous herborising resorts, to Surennes, Montmorency, St. Maur, Gentilly, and the islands in the Marne; how he was followed by young medical students from every land, from Danube, Thames and Tagus, Italians, Swedes and Irish, the latter full of zeal for their Prince and their religion, and bearing the lilies that are the arms of the Grand Monarque, the sole refuge of dethroned kings; and how Vaillant described each species and its medicinal uses and gently opened the buds with a pointed instrument to show the method of pollination.

A lengthy passage on "Borames," the Scythian lamb of the Caspian shores, is the chief foundation for the charge of plagiarism that has been brought against Erasmus Darwin by Isaac Disraeli and others. Darwin's *Loves of the Plants* was published in 1789, and in the first canto is the following reference to this interesting myth:

> " E'en round the pole the flames of Love aspire,
> And icy bosoms feel the secret fire!
> Cradled in snow and fan'd by arctic air
> Shines, gentle Barometz! thy golden hair;
> Rooted in earth each cloven hoof descends,
> And round and round her flexile neck she bends;
> Crops the grey coral moss, and hoary thyme,
> Or laps with rosy tongue the melting rime;
> Eyes with mute tenderness her distant dam,
> Or seems to bleat, a Vegetable Lamb."

Nothing more is known of MacEncroe, who refers to the bursting of a bog near Limerick. Although his poem was edited in 1791 by Sir Richard Clayton, the editor knew nothing of the author.

In the article " Generation " in his *Gardeners' and Florists' Dictionary* of 1724, Philip Miller makes the earliest known reference to insect pollination. He describes how, after removing the anthers from some Tulips, he saw a bee enter the flower and leave pollen on the stigma, which actually resulted in seed.

Linnæus mentions in his *Fundamenta Botanica* (1736) that in the garden in his father's parish of Stenbrohuld in 1723, when he himself was but sixteen, the male flowers of the Gourd were daily removed, with the result that no fruit was formed. In 1729, while at Upsala, " during this period of intense receptivity," as his biographer Dr. Jackson remarks, he came across a review of Vaillant's *Sermo* and a treatise by Wallin, published at Upsala, entitled *Nuptiæ Arborum Dissertatio*. Convinced of the importance of stamens and pistils, he determined to found a classification upon them, and drew up a tract entitled *Prœludia sponsalia plantarum*, which, though then shown to Celsius and Rudbeck, was not printed until 1908. Not until the work of Kölreuter, between 1761 and 1766, was it recognised that insects are among the most important agents in pollination. Nor was it until 1790 that Christian Konrad Sprengel (1750–1816) recognised that perfect flowers are by no means generally self-pollinating. He came to this conclusion by his observation of dichogamy, or the maturation of the anthers and stigmas of a perfect flower, such as *Epilobium angustifolium*, at different times. In 1799 Thomas Andrew Knight formulated, in the *Philosophical Transactions*, the doctrine that no plant self-fertilises itself for a perpetuity of generations. Not until 1830 did Amici trace the pollen-tubes from the pollen-grain into the ovary and the micropyle.

CHAPTER XXXI

SOME EARLY PHYSIOLOGICAL EXPERIMENTS

IT is at least as important that we should endeavour to know more of the life of the individual plant as it is that the vast variety of the vegetable productions of every corner of the globe should be described and classified. Workers in vegetable physiology have, however, often been ignored by the historians of botany. As we have seen, the crude experiments of Malpighi and the speculations of Grew were constantly at fault from their teleological preconceptions and reliance on the analogy with the animal world. On the other hand, if some of the main advances in physiology were made by men who were physicists rather than biologists, it was precisely because they were free from such preconceptions and approached the subject by exact experimental methods.

Edmé Mariotte, Prior of St. Martin sous Beaune, near Dijon, one of the greatest of physicists, was an original member of the Académie des Sciences founded in 1666, and died in 1684. In 1679 he propounded—in the form of a letter, " *Sur le sujet des plantes* "—conclusions of the greatest importance, as to plant-nutrition. In opposition to the Aristotelian view that plants took up by their roots the compounds required for their nutrition in an elaborated condition, he pointed out that poisonous and harmless plants grow side by side in the same soil, and that thousands of different kinds of plants may be grown in ordinary earth watered only by rain-water. He noticed also the high pressure under which the sap exists in plants, and concludes that this contributes to their growth; and it is interesting to find him protesting, as against the favourite sixteenth-century superstition known as the Doctrine of Signatures, that the medicinal properties of plants can only be ascertained by experiment, and not by resemblances in their external form to the organs of the human body.

While histology was in its infancy, and any real quantitative chemical science, especially that of gaseous substances, was practically non-existent, we can only be surprised at the results obtained by experimentalists such as Mariotte and Hales.

Stephen Hales was born at Bekesbourne, near Canterbury, in 1677. In 1697 he entered Corpus Christi College, Cambridge, where he graduated in 1700, and was made a Fellow in 1703, after which date he appears to have begun work in experimental chemistry. In 1709 he became perpetual curate of Teddington, and, although he was afterwards also Rector of Porlock in Somerset and (from 1722) of Faringdon, Hants, where he occasionally resided, Teddington was his home for the remainder of his life. Gilbert White of Selborne did not become curate of Faringdon till Hales's death in 1761, but it is pleasant to think that it may have been the advice of the experimentalist Rector (as it certainly was in the case of White's correspondent, Robert Marsham) that led to the keeping of those journals that subsequently grew into the *Natural History of Selborne*

Hales became a Fellow of the Royal Society in 1718, and his *Vegetable Staticks*, published in 1727, was originally communicated to the Society. Frederick, Prince of Wales, is said to have been fond of surprising him in his laboratory at Teddington. On the prince's death in 1751, Hales became Almoner and Clerk of the Closet to his widow, the Princess Augusta. When the Princess began the laying out of her botanical garden at Kew in 1761, Hales, who believed in the value of fresh air for plants as well as for human beings, was allowed to design the flues for heating the Great Stove. Incidentally it may be mentioned that this stove stood between the Temple of the Sun and the big *Wistaria* until 1861.

Writing of Hales to Marsham, White says: " His attention

to the inside of Ladies' tea-kettles, to observe how far they were incrusted with stone, that from thence he might judge of the salubrity of the water of their wells—his advising water to be showered down suspicious wells from the nozzle of a garden watering-pot in order to discharge damps, before men ventured to descend; his directing air-holes to be left in the out-walls of ground-rooms, to prevent the rotting of floors and joists . . . his teaching the house-wife to place an inverted teacup at the bottom of her pies and tarts to prevent the syrup from boiling over, and to preserve the juice . . . are a few, among many, of those benevolent and useful pursuits on which his mind was constantly bent.

" Though a man of a Baronet's family, and one of the best houses in Kent, yet was his humility so prevalent, that he did not disdain the lowest offices, provided they tended to the good of his fellow-creatures. The last act of benevolence in which I saw him employed was at his rectory at Faringdon, where I found him in the street with his paint-pot before him, and much busied in painting white, with his own hands, the tops of the foot-path posts, that his neighbours might not be injured by running against them in the dark. His whole mind seemed replete with experiment. . . . He used to lament to my Father, how tedious a task it was to convince men, that sweet air was better than foul, alluding to his ventilators. . . . It should not be forgotten that our friend was instrumental in procuring the Gin-act, and stopping that profusion of spiritous liquors which threatened to ruin the morals and constitutions of our common people."

In the preface to his *Vegetable Staticks* Hales regrets that Malpighi and Grew had not confined themselves to a " statical (or purely physical) way of inquiry." Although no doubt he was influenced by the work of Robert Boyle, he himself owes but little to his predecessors. His methods are both original and precise: he marks and measures growing shoots; uses a graduated column of mercury to determine the amount of root-pressure; is constantly weighing and re-weighing—as, for instance, when measuring loss by transpiration; and finds the area of leaf-surfaces by a net-work of threads forming a quarter-inch mesh. As Sachs says of him, he "may be said to have made his plants themselves speak, by means of cleverly contrived and skil-fully managed experiments . . . not content with giving a clear idea of the phenomena of vegetation, he sought to trace them back to mechanico-physical laws."

Hales's best-known experiments were the measurement of the transpiration from the leaf-surface of a Sunflower, and that of the root-pressure exhibited in the bleeding of a decapitated Vine-stem. Though he knows nothing either of stomata or of chlorophyllian action, and considers the leaves to be chiefly organs of transpiration to raise the sap from the root through the stem by suction, he yet recognised that in some way the air contributed to the building-up of the plant, as well as the liquid absorbed by the roots. He asks also: "May not light, which makes its way into the outer surfaces of leaves and flowers, contribute much to the refining of the substances in the plant?"

Although he actually obtained many different gaseous substances or "airs" in his experiments, he had no means of discriminating between them. It was not until half a century later that the theory of plant-nutrition took another step forward.

In 1779 Priestley found that the green parts of plants gave off the oxygen gas that he had discovered five years before; but the very title of the work of Ingen-Houss, pub-lished in the same year, shows that he had ascertained more of the rationale of the process.

Jan Ingen-Houss was born at Breda in 1730, graduated in medicine at Vienna, and came to practise in London in 1764. His first work, published in 1779, is entitled, "*Experiments on Vegetables, discovering their great power of purifying the common air in the sunshine and of injuring it in the shade and at night.*" Although the progress of chemistry during the next seventeen years enabled him to state his conclusions in more modern terms, it is clear that by 1779 he had discovered that all plants give out carbonic acid gas in respiration throughout life, and that only green leaves and shoots exhale oxygen, and that only in sunlight or clear daylight. This latter process he recognised as nutritive, and he saw that the carbon dioxide of the atmosphere was the chief, if not the sole, source of the carbon in the plant.

Limitations of space forbid us saying more about the later progress of physiological experiments in the history of English botany, and we must conclude with the mere mention of Thomas Andrew Knight (1759–1838). He was a country gentleman and one of the founders of the Royal Horticultural Society, and his experiments on the action of gravitation, by growing plants on a wheel driven by water-power, have become classical. A delightfully quaint story of Charles Daubeny (1795–1867), Professor of Botany at Oxford from 1834, must also be mentioned. He experimented upon the action of light on growth and used for one of his colour-screens Magdalen College port!

Q

CHAPTER XXXII

LINNÆUS

STOCK-TAKING is as useful in science as in business. If, as Gilbert White wrote, the whole mind of Hales seemed replete with experiment, that of Linnæus was certainly replete with classification. He classified not only plants but botanists, animals, minerals, and even diseases.

Carolus Linnæus (see frontispiece) was born on May 23, 1707, at Rashult, in Stenbrohuld, a district in the province of Småland in Southern Sweden. Here his father was minister, the family taking such names as Linnæus, Lindelius, and Tiliander from a tall linden tree on the farm at Rashult. The " Little Botanist," as he was called at school, inherited his passion for flowers from his father, who kept his garden gay with them and (so the story runs) used daily to decorate with garlands the cot of the future naturalist. At four years old the child is said to have been much impressed by his father's remarks about the uses of the plants of the neighbourhood. As he showed nothing but distaste, when he was eighteen, for those studies that were to fit him for the ministry, his father was recommended to apprentice him to a tailor or shoemaker. A physician of the town, however, assured the poor clergyman that his son might yet distinguish himself in medicine and natural history, and took the lad into his house, taught him physiology, and introduced him to Tournefort's *Institutiones rei herbariæ*, which had been published in 1700.

Linnæus' evident zeal during his youth in the constant acquisition of botanical knowledge impressed those who met him, and repeatedly secured for him kindly encouragement and assistance. In 1727 he proceeded to the university of

Lund, where he lodged with Stobæus, a physician and naturalist, who allowed him the use of his library and taught him how to preserve museum specimens. A year later he removed to Upsala, however, as being better suited for his medical studies, but here the eight pounds, which was all his father could give him, was soon spent. He could get no private pupils, had to mend his shoes with folded paper, and was often badly in want of a meal!

Celsius, the aged Professor of Theology, then engaged on his *Hierobotanicon*, a treatise on the plants of the Bible, finding him studying the plants in the University garden, took him into his house, made him free of his books and secured him some pupils. Then it was that, as we have already described, he came across the notice of Vaillant's *Sermo*, and under its influence he wrote his *Prœludia sponsalia plantarum*. Celsius having shown the manuscript of this treatise to Rudbeck, the aged Professor of Botany, the latter in 1730 made Linnæus his assistant and allowed him to lecture for him, to organise botanical excursions, and to remodel the botanical garden.

More than thirty years before Rudbeck, at Royal command, had made a botanical tour of Lapland, but his results had all been destroyed by fire in 1702. Now the Academy of Sciences at Upsala determined to revive this undertaking, and at a cost to them of about £25 Linnæus traversed in five months some 4600 miles of country, carrying fowling-piece, hanger, measuring rod, pressing-paper, pocket microscope, telescope, and but a small allowance of clothing. Though he only distinguished 537 specimens of plants, upwards of a hundred of these were new to Sweden. *Campanula serpylli-folia*, " a little northern plant, long overlooked, depressed, abject, flowering early," was obviously not a *Campanula*, and, requiring a new name, was chosen by Linnæus as appropriate to bear his own, and was accordingly described by Gronovius in the *Flora Lapponica* of 1737 as *Linnæa borealis*.

Whilst on a similar tour in Dalecarlia, in 1734, Linnæus

was strongly urged to graduate in medicine at some foreign university. In the following year he journeyed, by way of Lübeck and Hamburg, to Amsterdam, graduated M.D. at Harderwijk, and visited Haarlem and Leyden. Here he met Gronovius and Boerhaave. The former was so much struck by the manuscript *Systema Naturæ* that he printed it, in eight folio sheets, at his own expense. Boerhaave gave Linnæus an introduction to Burman at Amsterdam, with whom he stayed a year, during which he published his *Fundamenta Botanica*. The mere tabular outline of the *Systema* and the aphorisms of the *Fundamenta*—in which the terms *corolla, filament, monœcious*, and *diœcious* are employed for the first time—had an immense influence, the former passing through twelve, and the latter through five, other editions during the author's lifetime. Cliffort, a wealthy banker, then invited Linnæus to stay with him at Hartecamp, some three miles from Haarlem, to manage his large gardens, in which George Dionysius Ehret was then at work as an artist. In passing it may be remarked that Ehret, the son of a gardener at Durlach in Baden, was a born artist. His self-taught work had attracted the notice of the wealthy Dr. Trew of Nuremberg, who became subsequently the editor of Mrs. Blackwell's *Herbal*. Ehret had afterwards painted and taught painting at Basel, Montpellier, and Paris, where he had been employed by the de Jussieus. After illustrating the *Hortus Cliffortianus*, which Linnæus prepared, from living and dried specimens, in 1737, Ehret settled in London, married a sister-in-law of Philip Miller, illustrated Browne's *Natural History of Jamaica* (1756), and executed many beautiful paintings, often on vellum, for Sloane, Fothergill, the Duchess of Portland, and others.

At Cliffort's expense, Linnæus visited England in the spring of 1736, when the wild Hyacinths were in full bloom in the copses—a novel delight to his northern eyes. To Sloane, then seventy-six, Boerhaave gave him a letter of introduction, which ran:—" Linnæus, who will present you this letter, is well worthy to meet you and to be met by you:

whoever sees you together will see two men whose equal the world will scarcely provide."

The elderly President of the Royal Society seems not to have been too cordial to the young botanical revolutionary. He was, however, well received by Philip Miller, by John Martyn, and by the worthy Peter Collinson, the correspondent of Benjamin Franklin and of other American men of science. At Oxford, he writes years afterwards in his Diary:

" The learned Botanist, Dillenius, was at first haughty; conceiving Linnæus' *Genera*, which he had got half printed from Holland, to be written against him; but he afterwards detained him a month, without leaving Linnæus an hour to himself the whole day long; and at last took leave of him with tears in his eyes, after having given him the choice of living with him till his death, as the salary of the Professorship was sufficient for them both."

John Jacob Dillenius, a native of Darmstadt, had been brought to England by William Sherard. He had edited Ray's *Synopsis* in 1724; had been nominated by Sherard as, from the date of his own death (1728), first Sherardian Professor at Oxford; and had, in 1732, published the *Hortus Elthamensis*. He is best known, however, for his *Historia Muscorum*, published in 1741, which was the first systematic treatise on Mosses, and is illustrated by fine figures etched by himself. In August 1736, Dillenius wrote to a friend:

" A new botanist is arose in the north; a founder of a new method, ' *a staminibus et pistillis*,' whose name is Linnæus. . . . He . . . hath a thorough insight and knowledge of botany; but I am afraid his method won't hold. He came hither and stayed here about eight days, but is now returned to Leyden."

On his return to Hartecamp, Linnæus completed the printing of his *Genera Plantarum* (1737), the starting-point of modern systematic botany, and wrote the *Hortus Cliffortianus* and the *Critica Botanica*, which he dedicated to Dillenius.   The *Genera* contained diagnoses of 935 genera, which number was increased to 1336 in the five succeeding editions and two *Mantissæ* or supplements.   The *Critica* dealt in detail with the rules to be observed in naming genera and distinguishing species.

The damp air of the Netherlands seems to have ill suited Linnæus's northern constitution, bringing on attacks of ague, which were, perhaps, the forerunners of the gout of his later years.   Leaving Hartecamp, with the intention of returning to his native land, he stayed a year at Leyden, assisting in the rearrangement of the garden, and there produced his *Classes Plantarum* in 1738.   This is a history of the classification of plants from the system of Cæsalpinus in 1583 down to his own Sexual System of the *Systema* in 1735.   Its most original feature, however, is the acknowledgment that there is a natural system that could not be determined by any fixed set of selected marks, the rules for framing it being as yet undiscovered.   Sixty-five natural orders are enumerated, but no characters could be given for them.

By Cliffort's help, Linnæus then journeyed through the higher ground of Belgium to Paris, where his pleasure in meeting Bernard de Jussieu was lessened by his ignorance of French, for he had no aptitude for languages and always used Latin in speaking with foreigners.   From France he returned by sea to Sweden and established himself as a physician in Stockholm.

The death of Rudbeck in 1740, and of another professor soon after, placed Linnæus (in 1742, when he was thirty-five) in the position that had been the ambition of his life, the Professorship of Botany at Upsala.   He increased the plants in cultivation in the botanic garden by more than a thousand species, and the number of students in the university rose from five hundred to fifteen hundred.   Well-trained pupils of his travelled as naturalists into every unexplored region—Forskäl to Arabia, Hasselquist to Syria, Kalm to North America, Thunberg to South Africa, Japan and

Ceylon, and Solander on Cook's first voyage of circum-navigation.

In 1751 he published his *Philosophia Botanica*, containing a revision and expansion of the *Fundamenta*, *Bibliotheca*, *Classes* and *Critica* of earlier years. In 1753 he published his *Species Plantarum*, from which all the earliest plant-names now in use are taken. Rivinus in 1690 had proposed that no plant-name should consist of more than two words, but he left monotypic genera with but a single name. Linnæus was thus the first definitely to give to every plant a binary name. Some 7300 species are diagnosed in this work, with their synonymy and localities—arranged, of course, according to the Sexual System. Although the number is less than those described by Tournefort or Ray, almost all had been examined by Linnæus himself and were represented in his herbarium.

In the year of its publication Linnæus was created a Knight of the Polar Star, the first time that a man of science had been knighted in Sweden. In 1761 a patent of nobility was granted to him, from which date he was styled Carl von Linné. Although his practice as a physician had grown and he had become comparatively wealthy, his prodigious labours told early upon his physique. When he was sixty his memory began to fail, and, although he had warded off various attacks of gout by a diet of alpine Strawberries, he was seized by apoplexy, followed by paralysis, and died in 1778 at the age of seventy. He was buried in the cathedral of Upsala, sixteen doctors of medicine, all of whom had been his pupils, acting as pall-bearers.

We have by no means mentioned all the books that Linnæus produced, but it may be well to call attention to the volumes of graduation theses issued by him, under the title of *Amœnitates Academicæ*. Although nominally the work of the various candidates for degrees, these are in most cases—and certainly in the main—the work of Linnæus. They deal with the greatest variety of topics—for example, the nature of a drug or a disease; the plants of a district;

the terms used in descriptive botany (of which 673 are explained); the dispersal of seeds; the sleep of plants; or the methods of pollination. Although self-pollination is looked upon as the rule, and the anthers and stigmas in perfect flowers are believed to mature simultaneously, it is recognised that diclinous trees (those having stamens and carpels in separate flowers) commonly flower before leafing, so that wind-pollination is not obstructed; and that nectar serves as a food to pollinating insects.

The greatest services of Linnæus to botanical science were unquestionably the binominal nomenclature and the Sexual System of classification. The former has enormously simplified all reference to the descriptive work of others. While the number of known plants was rapidly increasing, a ready means of sorting the material was imperatively necessary, and Linnæus' method of twenty-four *classes* dependent on the number, union, and relative length of the stamens, subdivided into *orders* by the number of the styles, was the simplest ever devised. Linnæus himself, however, wrote to Haller: " I have never pretended that the method was natural," and he constantly taught that a natural system based upon " the simple symmetry of all the parts " must be the aim of the future. The Linnæan System was superseded, not because it was unscientific but from the impossibility of its universal application. It supposes a higher and more uniform development of flowers than has yet been attained. The Natural System is not absolute, but gives the nearest approach to a classification applicable to all plants.

Linnæus' only son succeeded him as Professor, but died in 1783, when his mother offered the collections and library to Sir Joseph Banks. They were, however, at his suggestion, purchased by Dr. (afterwards Sir) James Edward Smith, the founder and first President of the Linnæan Society of London, for £1000. At his death, in 1828, they were bought, by subscription among the Fellows, for presentation to the Linnæan Society, in whose reverential custody they remain to-day.

CHAPTER XXXIII

THE NATURAL SYSTEM IN FRANCE

FRANCE never accepted the Linnæan Sexual System. The sons of Montpellier carried on the tradition of Pena, Lobel, and Bauhin in striving after a natural arrangement of plants. Pierre Magnol (Plate XII), born in Montpellier in 1638, and Professor of Botany from 1694 till his death in 1715, seems to have anticipated Ray and Linnæus in recognising the futility of single characters as a guide to affinity. Whilst he is the first to employ (in his *Prodromus historiæ generalis plantarum, in qua familiæ per tabulas disponuntur*, 1689, 8vo) the happy term " family " for a natural group of plants, he merely names seventy-six families, without characterising them. In the Preface, however, he writes:—" Having examined the methods most in use and found that of Morison insufficient and very defective, and that of Ray much too difficult, I think I can perceive in plants a certain affinity between them, so that they might be ranged in divers families, as we class animals. This apparent analogy between animals and plants has induced me to arrange them in certain families, and, as it appeared to me impossible to draw the characters of these families from the fruit alone, I have selected the most striking characteristics I have met with, whether root, stem, flower, or seeds. There is among plants a certain similitude or affinity, as it were, which does not exist in any of the parts considered separately, but only as a whole."

In his *Novus character plantarum*, not published until 1720, he bases his primary divisions on the superiority or inferiority of the calyx, and his subdivisions on the corolla.

Magnol had among his pupils Tournefort and Bernard de Jussieu.

The Royal Garden of the Louvre (founded by Henri IV about 1590), which had so able a director as Jean Robin, the friend of Gerard, had been succeeded by the Jardin des Plantes, established in 1626 by Cardinal Richelieu. Here, in 1635, Vespasien son of Jean Robin, had planted the first Acacia (*Robinia*) in Europe; and here Tournefort was appointed Demonstrator in 1683.

Joseph Pitton de Tournefort (Plate XIII) was born at Aix in 1656. Destined by his father for the priesthood, Nature had made him a botanist, and he played truant from his seminary, bribed the keepers of enclosures to permit him to botanise, trespassed where he could not otherwise gain admission, and was nearly stoned to death by peasants, who took him for a thief. After his father's death he went (in 1679) to Montpellier, where he graduated in medicine, and explored the Cevennes, Pyrenees, and mountains of Catalonia. After his appointment to the Jardin des Plantes he was commissioned to travel in Andalusia, Portugal, England, and Holland in search of plants. When he published his *Élémens de botanique* (in 1694), which was illustrated by 450 fine copper-plates, he was able to include 10,146 species, under 698 genera. As Sachs says, he was esteemed by his contemporaries for this great knowledge of plants, though his classification evinces but little morphological insight. He expressly refused to admit the existence of sex in plants, and bases his twenty-two classes mainly on the corolla. With the primary division into trees and herbs, each subdivided into petaliferous and apetalous, he divides petaliferous herbs into those with simple or compound flowers, and petaliferous trees and simple flowers alike into monopetalous and polypetalous groups, these being again divided into regular and irregular. Apart from the separation of arborescent *Rosaceæ* and *Papilionaceæ* from the herbaceous members of those families, this gives a good many natural groups in the twenty-two classes. Experience

PIERRE MAGNOL (1638-1715).

facing p. 240.

Plate XI.

proved this grouping to be not only artificial but inadequate to the needs of enlarging knowledge.

In contradistinction to the plan adopted in Bauhin's *Pinax*, in which the genera are merely named but the species are diagnosed, Tournefort characterises his genera, but not his species. In France, however, this system prevailed for nearly a century—until, in fact, it was superseded by that of de Jussieu. In England its simplicity recommended it until the Linnæan system prevailed, *i.e.* about 1760.

Tournefort published a Flora of the environs of Paris in 1698 and an enlarged edition of the *Élémens* in Latin, under the title of *Institutiones rei herbariæ*, in 1700, in which year he started on a three-years' journey to the Levant. His death singularly parallels that of his great British predecessor Morison, for in 1708 he was struck in the chest by the pole of a carriage while crossing the street near the Jardin des Plantes, and died a month later. The journal of his last tour was published after his death. It contains an account of his ascent of Mount Ararat, but includes none of that description of successive altitudinal zones of vegetation ascribed to him by Humboldt, apparently by a *lapsus memoriæ*, and copied by many subsequent writers.

In the history of botany there are not a few cases of the hereditary succession of ability through two or more genera-tions. Of this a striking example is that of the family of de Jussieu, of which no less than five members contributed to the progress of the science. Antoine (1686–1758), Bernard * (1699–1777), and Joseph (1704–79) were three sons of an apothecary at Lyons. The two elder brothers studied under Magnol at Montpellier and travelled together in Spain and Portugal. Antoine succeeded Tournefort in 1708, but did little original work during his half-century of office. Joseph accompanied La Condamine's scientific mission to Peru and remained thirty-six years in South America, whence he sent to his brother Bernard the first seeds of *Heliotropium peruvianum*. Bernard was summoned to Paris by his elder

* See Plate XIV.

brother to succeed Vaillant as Sub-demonstrator at the Jardin des Plantes, one of his chief duties being to superintend the herborisations of the students. There is a pretty story of a Frenchman in the Holy Land finding a seedling Cedar at Lebanon and bringing it with tender care to France, sharing with it his scanty drinking water on the voyage, and carrying it in his hat. The Frenchman was Bernard de Jussieu, but the Holy Land was England, and Lebanon probably Sherard's garden at Eltham, for de Jussieu came to England for plants in 1726. He visited Eltham, and it was Sherard who gave him the seedling, which, after flourishing for more than a century in the Jardin des Plantes, was sacrificed to a railway extension.

In 1759 Bernard was invited to arrange the garden of La Trianon at Versailles for botanical study, and decided to do so on the lines of the *Classes Plantarum* and *Philosophia Botanica* of Linnæus, with whom he had had much friendly intercourse in 1738. Taking as his starting-point Ray's division of Dicotyledons from Monocotyledons, he used the superiority or inferiority of the ovary and the union or absence of union among the petals as the characters of next importance, separating diclinous and apetalous plants and conifers.

A man of singular modesty, Bernard de Jussieu published nothing of his own, and on his brother Antoine's death refused to succeed him. In 1765, growing old and infirm, he summoned a nephew, Antoine Laurent de Jussieu (Plate XV) (1748–1836), to assist him, and to his lot it fell to enunciate and expound his uncle's system.

In 1773 this young man communicated to the Académie des Sciences a memoir on the *Ranunculaceæ*, which has been called the birth of the Natural System. In it he enunciated the principle of the relative value of characters—that, as he put it, characters must be weighed, not counted. This *Examen de la Famille des Renoncules* was followed in the next year by his *Exposition d'un nouvel ordre de Plantes*, in which he extended his system to the other families. His primary division is into Acotyledons, Monocotyledons and Di-

JOSEPH PITTON DE TOURNEFORT (1656-1708).

facing p. 242

Plate XII.

cotyledons, these last being subdivided into Apetalæ, Monopetalæ, Polypetalæ, and Diclinæ, while the further subdivisions are based upon the insertion of the stamens, whether hypogynous, perigynous, epigynous or epipetalous —a character the importance of which was then first emphasised.

The Trianon garden was rearranged, and after fifteen years' more hard work at the elaboration of the system, and (on July 14, 1789) at a time when all Paris was crying: "Down with the Bastille" and, later, "The Bastille is taken," there issued from the press the last sheet of his *Genera plantarum secundum ordines naturales dispositæ, juxta methodum in horto regio Parisiensi exaratum anno* 1774. In this he makes fifteen classes comprising 100 orders and 1754 genera, and, with the exception of one or two little blunders—such as the placing of *Hippuris* with *Chara* among the Acotyledons and *Nymphæa* among Monocotyledons— these orders are mostly retained to-day. In many cases their sequence is also that which we now consider to represent their affinities.

This great work was followed by a long and valuable series of monographs on different families in the *Mémoires* of the Museum of Natural History that de Jussieu organised in 1793. In 1826 he resigned his professorship to his son Adrien (1797–1853), who is best known for his text-book entitled *Botanique*, translated into most of the principal languages of Europe.

Amidst an enormous mass of other valuable botanical work, Augustin Pyrame de Candolle (1778–1841)—a member of a Provençal family, settled in Geneva—did much to establish sound principles of classification. From 1808 to 1816 he was Professor of Botany at Montpellier, and from 1817 in his native city. The secondary title of his *Théorie Élémentaire de la botanique*, published in 1813, is *Exposition des principes de la classification naturelle*, and in it he lays down the essential rule that anatomy must be the sole basis of classification to the exclusion of physiology. It was

in the second edition of this work (1819) that the term
" carpel " was first employed. In the last edition, that of
1844, edited by his son Alphonse, 213 " orders " (or, as we
now term them, families) are described.

Thus, while we are indebted to Caspar Bauhin for the
first distinct diagnoses of species, and to Tournefort for the
first of genera, it is to Antoine Laurent de Jussieu that we
owe the first characters of natural families. From French-
speaking botanists we have that series of progressively
enlarged circles of affinity that represent the growth of our
knowledge from the individual to the general.

BERNARD DE JUSSIEU (1699-1777).

facing p. 244.

Plate XIII.

CHAPTER XXXIV

KEW

It is just five hundred years since Henry V—to atone for his father's murder of Richard II—established, on each side of the Thames, the Bridgettine Nunnery of Syon and the Carthusian Priory of Sheen. At the latter, Cardinal Pole was educated, and Dean Colet, founder of St. Paul's School, built himself a Lodge. As we have already seen, it was at Syon House after the Dissolution of religious houses, that William Turner acted as a physician to the Protector Somerset. He had also a garden of his own somewhere at Kew, which is, perhaps, the Quay-hoe, or landing-place, of the ferry from Brentford. In 1678 Evelyn writes in his *Diary* :—" I din'd at Mr. Hen. Brouncker's, at the Abbey of Sheene, formerly a monastery of Carthusians. . . . Within this ample enclosure are several pretty villas and fine gardens of the most excellent fruites, especially Sir William Temple's (lately Ambassador into Holland). . . . Hence I went to my worthy friend Sir Henry Capel at Kew, brother to the Earle of Essex. It is an old timber house, but his garden has the choicest fruit of any plantation in England, as he is the most industrious and understanding in it."

Hester Johnson, Swift's " Stella," was born at Sheen; and there Jonathan Swift first acted as Temple's secretary. About 1707 the Lodge, which was built by Colet, was granted to the Duke of Ormonde, who rebuilt it. In 1721 it became the property of the Prince of Wales, afterwards George II, whose wife, Caroline of Anspach, became much attached to it and lavished vast sums of money on it and on the grounds. The gardens extended over much of what is now the Old Deer Park, Richmond, and the Arboretum at Kew, between the Thames and a bridle-path, called Love Lane, that ran from Richmond to the Brentford Ferry, and which is now represented by the Holly Walk in Kew Gardens.

In 1730, Frederic, Prince of Wales, took a lease of the house that had been Sir Henry Capel's at Kew. The grounds extended on the east side of Love Lane, whilst those of his father were on the west of it. Here it was that his widow, the Princess Augusta of Saxe-Gotha, with the advice of Lord Bute, decided to establish a botanical garden.

Bute (1713–92) was an enthusiastic botanist, who collected a library of 300 folio volumes on the subject. In 1785 he produced a sumptuous work entitled *Botanical Tables, containing the different Familys of British Plants, distinguished by a few obvious parts of fructification.* It consists of nine quarto volumes, the letter-press engraved and the plates coloured. Only twelve copies were printed, and it is said to have cost £12,000. In 1787 he also printed privately a *Tabular Distribution of British Plants.* He seems also to have planned the collection of plants from all quarters of the globe, for as early as 1750 a letter states that " the Prince of Wales is now about preparation for building a stove three hundred feet in length; and my Lord Bute has already settled a correspondence in Asia, Africa, America, Europe, and everywhere he can."

In 1759 William Aiton (1731–93), trained at Chelsea under Philip Miller, was placed in charge of the garden (which then consisted of only nine acres). Sir William Chambers, the architect of Somerset House, was commissioned to build the Orangery, now the Timber Museum. Within the three next years Chambers erected the Pagoda, most of the little temples with which, according to the taste of the time, the gardens are studded, and the Great Stove (pulled down a century later), for which, as we have seen, Hales designed the hot-air flues. It was 114 ft. in length and larger than any other then in the country. In 1762

ANTOINE LAURENT DE JUSSIEU (1748-1836)

facing p. 246.

Plate XIV.

Archibald, Duke of Argyll, whom Horace Walpole dubbed
" the tree-monger," gave the Princess a number of trees
from his garden at Whitton, near Hounslow, and of these a
fine Cedar, a Persimmon, a *Robinia*, and a Turkey Oak are
living to-day.

During the first years of his reign, George III and Queen
Charlotte, who became as much attached to botany as was
her mother-in-law, lived occasionally at Richmond Lodge.
There, in 1768, the Kew Observatory was erected by
Chambers for the observation of the transit of Venus.
About 1770, however, the King pulled down the house and
the adjoining hamlet of West Sheen, turning what is now the
Old Deer Park into pasture. When, in 1772, his mother
died, George came to live at Kew for the summer months,
and the Richmond Gardens and Arboretum were ultimately
united to his mother's. In the meantime, Lord Bute had
retired to his country house at Highcliffe, near Christchurch,
where his death is said to have been caused by injuries
received in climbing a cliff to reach a rare plant. He was
replaced as royal adviser by Sir Joseph Banks, who practically acted as an unpaid Director of Kew Gardens until
his death in 1820.

Banks, though belonging to a family with estates in
Lincolnshire, was born in London in 1743. Educated at
Harrow, Eton, and Oxford, he had made up for the
deficiencies in the University teaching by procuring a
botanical tutor (Israel Lyons) from Cambridge at his own
expense. As we have already mentioned, he had also
studied in the Chelsea Garden. In 1766–7 he went on a
voyage to Newfoundland, and from 1768 to 1771 accompanied Cook on his first great voyage of circumnavigation,
at a cost to himself of £10,000. He took with him Daniel
Charles Solander (1736–82), a favourite pupil of Linnæus,
who had come to England in 1760, was assistant-librarian
in the British Museum and afterwards acted as secretary
and librarian to Banks. He also took Sydney Parkinson
and two other artists, but these three unfortunately died

R

on the voyage. Parkinson made nearly a thousand drawings, and the more interesting plants were roughly described by Solander at the time. Those drawings of the plants of Tierra del Fuego, St. Helena, Otaheite and the Pacific, New Zealand, New Holland, Java, Madeira and Rio, are now preserved in the Botanical Department of the British Museum, each in a separate volume, some of them fully prepared for the printer. On their return the collections were housed at Banks's London residence in Soho Square, and botanists were freely permitted to examine them. Banks superintended the engraving of the drawings on copper, and Solander worked out the descriptions of most of the plants.

In 1772, as they were unable to accompany Cook on his second voyage, Banks and Solander went to Iceland and the Scottish islands. In 1773 Solander was made Keeper of the Printed Books in the British Museum, and in 1778 Banks became President of the Royal Society. In that year he wrote to the younger Linnæus that 550—or about half—of the drawings were engraved, and his correspondent dedicated the genus *Banksia* in his honour with reference to the great work then expected from the two collaborators. In 1782 Banks tells another correspondent that nearly 700 plates were engraved, but in that year Solander died from apoplexy at the early age of forty-six, and the great work never appeared. Banks's Journal was edited and published by Sir Joseph Hooker in 1896, and 328 of the copper-plates of Australian plants were lithographed and published in 1900-5, with Solander's descriptions critically edited by Mr. James Britten, for the Trustees of the British Museum.

Banks's energetic management of Kew doubtless occupied much of his leisure, during which time his suburban residence was at Spring Grove, Isleworth.

In 1772, Francis Masson, a native of Aberdeen, trained as a gardener under Aiton, was sent out as the first of the series of Kew collectors who introduced an immense number of new plants into cultivation. He twice explored South

Africa—on the second occasion (1786) in company with
Thunberg, the pupil and successor of Linnæus—and it is to
him that our gardens are indebted for most of the Pelar-
goniums, Heaths, and bulbous plants from that region.
Between 1778 and 1782 he collected in Madeira, the
Canaries, the Azores, and the West Indies, where he was
taken prisoner by the French and lost part of his collections.
Between 1783 and 1785 he was in Portugal and Spain, and
in 1797, after being again captured by a French privateer,
he reached North America. Here he collected until
December 1805, when he succumbed to the cold winter at
Montreal.

Another Kew gardener, David Nelson, accompanied
Captain Cook on his third voyage (1776–9), bringing back
the first specimen of *Eucalyptus*. In 1787 he was attached
to H.M.S. *Bounty*, under Captain Bligh, who had been, at
Banks's recommendation, commissioned to introduce the
Breadfruit (*Artocarpus*) from the South Sea Islands to our
West Indian possessions. In the historic mutiny that
followed, Nelson was cast adrift with Bligh, and, after a
journey of 3600 miles in an open boat, reached Timor only
to die. The successful introduction of the Breadfruit into
Jamaica was, however, accomplished in 1791 by another
Kew gardener, Christopher Smith.

Meanwhile, after many years' preparation, there appeared
(in 1789) the three octavo volumes of the first edition of the
Hortus Kewensis, which was issued by William Aiton. It
contained descriptions of some 5600 exotic plants that had
been introduced into Kew and other London gardens. The
descriptions were arranged on the Linnæan system, and
many had been the work of Solander.

After Solander's death (in 1782) Banks secured as his
librarian another Swedish pupil of Linnæus, Jonas Dryander
(1748–1810), an able botanist with unequalled biblio-
graphical knowledge and industry. Dryander was mainly
responsible for seeing this *Hortus Kewensis* through the press,
and it was he who described the Bird-of-Paradise flower,

dedicated to Queen Charlotte under the name of *Strelitzia Regina*, another species, discovered by Masson at the Cape, being named *S. Augusta*, after the King's mother. As the real authors' names do not, as a rule, appear in the book, the names have to be quoted as of Aiton, though he does not seem to have been much of a botanist.

Aiton died in 1793 and was succeeded by his son William Townsend Aiton (1766–1849), who was, perhaps, more of a botanist. In the preparation of the second edition of the *Hortus Kewensis* (5 vols., 1810–13) he depended much upon Dryander for the first two volumes; and, after his death in 1810, similarly upon Robert Brown.

The able and munificent support that Banks gave to Botany is nowhere more strikingly seen than in his securing the aid of Robert Brown and the two Austrian brothers, the inimitable botanical draughtsmen, Franz Andreas and Ferdinand Lucas Bauer. In 1784 the younger brother, Ferdinand (1760–1826), had accompanied Professor John Sibthorp of Oxford to Greece and prepared for his posthumous *Flora Græca* the drawings now preserved at Oxford. In 1801 he accompanied Flinders's expedition to Australia, to which Brown was attached as naturalist. After his return he published *Illustrationes Floræ Novæ Hollandiæ* (1813), besides illustrating many other important works. Franz Bauer, the elder brother (1758–1840), was engaged in 1790, at Banks's expense, as official draughtsman at Kew, and there he remained until his death, receiving an annuity after the death of his patron.

When George III became permanently insane (in 1810), Kew received but little Government support. Banks thought it best, therefore, to bequeath his extensive herbarium and library, with his house in Soho Square, to his librarian, Robert Brown, for his life with a reversion to the British Museum. This great collection, including Solander's manuscripts and Bauer's drawings, added to the Sloane Herbarium, became the nucleus of the Botanical Department of the Museum, of which Brown, in 1827, became the first Keeper.

Banks died in the same year as George III (1820), and the decline of Kew became even more marked. At last, in 1828, the Government contemplated its entire conversion into a Royal fruit and vegetable garden. A protest was raised, however, largely through the instrumentality of George Bentham, whose subsequent benefactions to Kew outshone even the services of Banks. A Treasury committee issued a report, drawn up by John Lindley, recommending that Kew should be made a National Garden, as a centre for all similar establishments in the Empire, and to advise Government in the founding of colonies, and foretelling that such a scheme would redound to the advantage of medicine, commerce, agriculture, horticulture, and many branches of manufacture.

One of the last works of Sir Joseph Banks—twenty years before—had been to secure the Regius Professorship of Botany at Glasgow for a promising but inexperienced botanist, William Jackson Hooker. Having proved his great ability, he was now appointed (in 1841) Director of Kew, on the resignation of W. T. Aiton. From that moment the history of Kew has been one of steady growth in area, in activity, and in utility to the interests of the nation and of botanical science.

In 1848 the first Museum of Economic Botany was established, the collections formed by Hooker when at Glasgow being its starting-point. In 1853 the Herbarium, now by far the most extensive in the world, was begun by the gift to the nation of the plants and books collected by Dr. Bromfield. To this were soon added a far larger library and herbarium presented by Bentham, and (in 1866) the even more extensive collections made by Sir W. J. Hooker, which were purchased by Government. The collection is now estimated to contain over two million specimens, and the library comprises 24,000 volumes.

In 1856 Hooker inaugurated the systematic preparation under Government auspices of a series of Floras of all the British Colonies. Those of Hong Kong and Australia were mainly the work of Bentham; that of the West Indies, of Grisebach; while that of India (1872-97) was largely the work of Sir Joseph Hooker, who acted as his father's assistant-director from 1855 until he succeeded him in 1865. That of Ceylon was left unfinished by Henry Trimen on his death in 1896, and finished by Sir Joseph Hooker between 1898 and 1900; that of South Africa was edited by Sir William Thiselton-Dyer, who acted as Director from 1885 to 1905; and that of Tropical Africa by Dr. Daniel Oliver.

Works of even greater import to botanical science have been the *Genera Plantarum* of Bentham and the younger Hooker (3 vols., 1865-83), and the *Index Kewensis*. From 1854, Bentham, though living in London, worked hard at Kew from ten to four daily, and a stupendous amount of systematic work was the result. The *Genera* contained diagnoses of 7569 genera of Phanerogams arranged under 200 " orders," in a system based upon that of Jussieu, and representing over 97,000 species. The *Index Kewensis* was published from Kew between 1893 and 1895 in four large quarto volumes of 2500 pages in all, with about 375,000 names. As all post-Linnæan synonyms are supposed to be included, this number gives no indication of the number of recognised species of Phanerogams. The work, rendered possible by a bequest from Charles Darwin, was carried out by Dr. B. Daydon Jackson, Sir Joseph Hooker, then upwards of eighty years of age, reading all the proofs.

Work of less scientific importance, but of the greatest utility to our Empire, has been the transfer of plants of great economic value, foods, drugs, dyes, timber, fibres, etc., from their native countries to those of our dominions in which they will grow. Of the Breadfruit we have already spoken. Tea was introduced about 1851 from China to Assam, by Robert Fortune, a collector sent out by the Royal

Horticultural Society. Kew took a leading part in two
other important transfers, the introduction of *Cinchona* from
the Andes to India, and that of Para Rubber (*Hevea brasiliensis*) from Brazil to Ceylon and the Malay Peninsula,
about 1860 and 1875 respectively.

The late Mr. Chamberlain, when Colonial Secretary, said
publicly:—" I do not think it too much to say that at the
present time there are several of our important Colonies
that owe whatever prosperity they possess to the knowledge
and experience of, and the assistance given by, the authorities
at Kew Gardens."

CHAPTER XXXV

PHYSIC: FARCES AND FEUDS

REFERENCES in John Martyn's *Virgil* (1740), Dillenius's *Historia Muscorum* (1741), and in other works showed that the writings of Linnæus were making an impression on English botanists within a few years of his visit to this country. In 1754 a *Flora Anglica*—consisting of a list of the Flowering plants in Dillenius's edition of Ray's *Synopsis*, with binominal names arranged under the Sexual System—prepared by Isaac Olaus Grufberg, a pupil of Linnæus, appeared at Upsala and was included in the *Amœnitates Academicæ*.

About this time that " learned and excellent " tea-party lion Benjamin Stillingfleet (1702–71), whose blue stockings became proverbial, turned his attention to botany, and the Linnæan system was probably made known to him by Robert Marsham, the friend of Hales and Gilbert White. In 1759 Stillingfleet published a translation of six essays from the *Amœnitates* with a preface that has been styled " the first fundamental treatise on the principles " of Linnæus published in England. It was Stillingfleet who urged the apothecary William Hudson to prepare his *Flora Anglica*, first published in 1762, and probably assisted him in the work.

Hudson (1730–93), a native of Kendal, was in 1757 and 1758 a sub-librarian at the British Museum, a fact that gave him access to the treasures of the Sloane Herbarium. He practised in Panton Street, Haymarket (Stillingfleet lived close by in Panton Square), and acted as " Præfectus " of the Chelsea Garden from 1765 to 1771. In 1783 his house was destroyed by fire, however, and he removed to Jermyn Street, where he died in 1793. His herbarium, bequeathed to the Society of Apothecaries, is now in the Botanical Department of the Natural History Museum, South Kensington, and many of the sheets show signs of the fire of 1783. Hudson's work, of which a second edition appeared in 1778, is the authority for the names of many British plants, and for many years it took the place of Ray's *Synopsis* as the annual prize volume for the best student at Chelsea, and as the field companion of most British botanists.

The year 1760 was notable in the history of the Linnæan system, for not only did Solander arrive in England in that year, but it saw also the publication of James Lee's *Introduction to Botany* and of John Hill's *Flora Britannica*.

Lee (1715–95) was one of the leading nurserymen of his time, having a garden at Hammersmith and sending out his own collectors to America and South Africa. It is to him that we are indebted for the Chilian *Fuchsia coccinea*. His *Introduction* is mainly a translation from Linnæus's *Philosophia Botanica*, which was carried out by Samuel Gray, father of the author of the *Natural Arrangement of British Plants*. Its popularity is shown by its going into nine editions within the succeeding fifty years.

As for Hill and his work, they belong to an altogether different class. He seems to have been a genius, combining the humdrum " unlimited capacity for taking pains " (Johnson's definition of genius) with the love of self-advertisement and absence of much of the troublesome restraint that we call conscience, which has also too often characterised " geniuses." The son of a Lincolnshire clergyman, he was born about 1716. Having been apprenticed to an apothecary, he started business for himself on the humblest scale, married a penniless girl and learnt what it was to be without the bare necessities of life. He travelled about the country collecting plants, and these he dried and made up into sets with descriptions. At one time he tried his fortune on the stage. In 1746 he was successful with an annotated edition and translation of Theophrastus's *History of Stones*, and from that time onwards his works, including magazines, farces, novels, and botanical works, seemed to pour from the press. In 1751 he began to write a daily letter, called the *Inspector*, to the *London Advertiser and Literary Gazette*, for which he collected scandalous gossip at the coffee-houses. He was always to be seen in the front row at the theatre and attracted attention by his magnificent attire. Cultivating " the gentle art of making enemies," he published a severe, though not unfounded, *Review of the Works of the Royal Society*, and also had a controversy with Garrick, who wrote in reply:

> " For physic or farces his equal there scarce is ;
> His farces are physic, his physic a farce is."

Hill purchased the degree of M.D. from the University of St. Andrews in 1750, and offered for sale various herb-medicines, such as " essence of water-dock " and " tincture of bardana," obtaining his herbs at first from the Chelsea Garden. When the scale of his gatherings led to his being refused admission, he started a garden of his own at Bays-water, near the present site of Lancaster Gate, and made large sums both by the *Inspector* and by his quack nostrums. In 1751 he published a folio *History of Plants*, mainly compiled from Linnæus, but suffering from the fact that the *Species Plantarum* had not then appeared. On the other hand, his *British Herbal* of 1756, also in folio, is a work of consider-able originality. It is our earliest post-Linnæan authority for *Valerianella*, *Linaria*, *Nymphoides*, etc., and—although the generic names are often borrowed from Tournefort, and when monotypic [1] have, as in Rivinus, no specific names— it is evident that he generally knew the plants of which he was writing. He objects to the artificial character of the Linnæan system, and employs one of his own in which 35 " classes " (two cryptogamic) are based upon characters mainly derived from the corolla and gynæceum. In the following year he published an essay on *The Sleep of Plants and Causes of Motion in the Sensitive Plant*, in which the alterna-tion of light and darkness is recognised as the efficient cause.

[1] *i. e.*, represented only by a single species.

Although containing original experiments on *Mimosa* and *Abrus*, this was, no doubt, suggested by an essay in the *Amœnitates*, the substance of which had been published the same year in the *Gentleman's Magazine*.

In 1758 Hill also issued *The Gardener's New Kalendar* (with) *The System of Linnæus also explained*. About this time he seems to have secured the patronage of Bute, which was not much to his ultimate advantage. At Bute's suggestion he began the issue of the *Vegetable System* in twenty-six folio volumes, with 1600 plates. This was published, between 1759 and 1775, at 38 guineas plain and 160 guineas coloured! After his death, his widow asserted that Bute had guaranteed him against loss for this undertaking and that he had failed to keep his promise. The scheme of the book is as lavish as its execution, though it has been cruelly dismissed as "cumbrous and useless."

The *Flora Britannica* of 1760 is professedly merely a re-arrangement of Dillenius's edition of Ray's *Synopsis* according to the Linnæan system. It has no alteration of nomenclature, so that it was naturally at once superseded by Hudson's work. In that year, however, Hill was—probably by Bute's influence—made gardener at Kensington Palace, a post worth £2000 per annum, and he seems to have been also nominated as Superintendent of the Royal Gardens, Kew—perhaps before Aiton's appointment. In 1768 he published a *Hortus Kewensis* in which it was stated that there were already more than 3400 species in the garden. In 1770 he published his interesting little octavo volume on *The Construction of Timber*. Sending a copy of his *Vegetable System* to the King of Sweden, he was made Knight of the Polar Star in 1774. He promptly used the title of Sir John Hill, but did not live long to enjoy the "honour," dying in 1775, heavily in debt.

The Linnæan system was now becoming firmly established. Alston, one of the last opponents of the existence of sex in plants, died in 1760. He was succeeded as Professor of Botany at Edinburgh by John Hope, who, although educated by Alston and by Bernard de Jussieu, became an ardent Linnæan. Himself a practical physiological experimentalist, Hope encouraged field botany and, by the pupils he trained, proved himself to be one of the greatest of botanical teachers.

In 1775, the year of Hill's death, Hugh Rose, an apothecary of Norwich, published a translation of the *Philosophia Botanica* under the title of *Elements of Botany*. William Withering, one of Hope's early pupils and then a successful physician at Birmingham, brought out the first edition of his *Botanical Arrangement of all the Vegetables naturally growing in Great Britain, according to the System of the celebrated Linnæus; with an easy Introduction to the Study of Botany*. Withering was the intimate friend of Priestley and a member of the so-called Lunar Society, riding to one another's houses on the Monday nearest to the full moon. To this belonged James Watt the engineer and Erasmus Darwin, whose *Loves of the Plants*, which forms part of his *Botanic Garden*, we have already quoted. Erasmus Darwin, grandfather of the great nineteenth-century naturalist, edited translations of Linnæus's *Systema Vegetabilium* and *Genera plantarum*. This appeared as *A System of Vegetables* (1783, 2 vols.) and *The Families of Plants* (1787, 2 vols.) " by a Botanical Society of Lichfield," although he, Sir Brooke Boothby, and a self-taught printer named Jackson, who was the actual translator, were the only members of the so-called Society.

It was curiously, as he himself tells us, on the day of Linnæus's death (January 10, 1778) that a lad of eighteen, who was to play an important part in the maintenance of the Linnæan system in England, began the study of botany. This was James Edward Smith (1759–1828), eldest son of a wealthy Norwich wool-merchant. Assisted in his early studies by Hugh Rose, and a group of Norwich botanists, correspondents of Stillingfleet and Hudson, Smith went (in 1781) to study medicine at Edinburgh. Here he studied botany under Hope, and made the acquaintance of Broussonet and of Richard Antony Markham, afterwards

Salisbury. In September 1783, Smith came to London to study anatomy under John Hunter, bringing with him an introduction from Hope to Banks, and he happened to be breakfasting with Banks on December 23 when the letter arrived from the executor of the younger Linnæus offering the entire collections and library for a thousand guineas. Banks declined the offer and urged Smith to accept it, and this, with his father's assistance, he did. In the following autumn the collections were despatched to England. It is stated that when Gustavus III of Sweden, who had been absent in France, heard of the sale he sent a vessel to the Sound to intercept the ship conveying them; but it was too late. Professor Sibthorp of Oxford had also been only just too late with a bid for the collection. Smith took apartments to house his treasure in Paradise Row, Chelsea, so that Chelsea became for a short time once more the Mecca of botanists, as it had been when Sloane's collections were at the Manor House. Thither came Banks and Dryander during the winter of 1784–5 to go through the herbarium, and assist with the unpacking of the rest of the collection.

Smith was elected a Fellow of the Royal Society in 1785, and in the following year took his medical degree at Leyden, where he inspected a specimen of the South European Fanpalm, *Chamærops humilis*, planted by Clusius two hundred years before. He then travelled through France, Italy, and Switzerland, his possession of the Linnæan collections, and letters from Banks, securing him a cordial reception everywhere. On his return he founded the Linnean Society, Banks, Salisbury, Sibthorp, Dryander, and Thomas Martyn being among the thirty-six original Fellows. The first meeting was held on April 8, 1788, at the house he had then taken in Great Marlborough Street.

Smith's more important works were *English Botany*, in 36 volumes 8vo., 1790–1814, often called Sowerby's English Botany, from the name of the publisher, who was also the draughtsman of the 2592 fine coloured plates; *Flora Britannica*, 3 vols. 8vo., 1800–4; the *Prodromus Floræ Græcæ*, 2 vols.

8vo., 306–13, and six volumes of the folio edition of his friend Sibthorp's posthumous *Flora Græca*; the *Introduction to Physiological and Systematic Botany*, 1807, which passed through six editions during his lifetime; the biographies of botanists and numerous other articles contributed to Rees's *Cyclopædia*, 1808–19; and *The English Flora*, 4 vols. 8vo., 1824–8. His fluent style did much to popularise botany, but he seemed to think that his possession of the Linnæan collections gave him the right to lay down the law in all matters botanical, while—more Linnæan than Linnæus—he would not hear of the supersession of the artificial Sexual System by one more natural. Although after his marriage in 1796 he was seldom in London, he was annually re-elected President of the Linnean Society until his death in 1828.

As we have seen, Richard Anthony Markham had been Smith's fellow-student at Edinburgh. Born at Leeds in 1761, he is said to have been descended on his mother's side from Henry Lyte, the translator of Dodoens. He was early left a considerable fortune on taking the name of Salisbury, and Smith, in the early days of their friendship, dedicated to him the name *Salisburia adiantifolia*. This in his purism he substituted for the Linnæan *Ginkgo biloba* as the appellation of that remarkable survival from earlier geological times known as the Maiden-hair tree. Salisbury had a fine garden at Chapel Allerton near Leeds, and in 1802 purchased the estate at Mill Hill, which had been the home of Peter Collinson. Here he is believed to have brought Linnæus in 1736, and many fine trees in the grounds of the famous school still commemorate its former owners. Smith visited Salisbury at Mill Hill, but the latter was far too good a botanist to accept the Linnæan system as final, and said so with a frankness that offended Smith.

In 1806 William Hooker (1779–1832), whose name is still borne by a shade of green popular with artists, an apt pupil of Franz Bauer, began the issue of a series of plates under the title of *Paradisus Londinensis*, for which Salisbury supplied the descriptions, and, perhaps, part of the funds.

In its pages, in March 1808, Salisbury described a genus of
Californian Hyacinths under the name *Hookera*. A little
later Smith described a genus of Mosses under the name
Hookeria, commemorating his " young friend, Mr. William
Jackson Hooker of Norwich, F.L.S." He re-described the
Californian plants under the name *Brodiæa*, after his friend
James Brodie of Brodie. If, as seems evident, there was a
conspiracy between Smith, Banks, Brown, and others thus to
ignore Salisbury's work, because they disliked the man,
Salisbury seems to have been quite capable of retaliating in
kind. In January 1809, Brown read a paper at the Linnean
Society on the *Proteaceæ*, which was not published for two
years. Salisbury was present at the reading of the paper.
Shortly afterwards a volume on the Order appeared,
nominally by Joseph Knight (a gardener who had established the business in King's Road, Chelsea, long carried
on by the Messrs. Veitch), but obviously the work of Salisbury. In this, many of the plants described in Brown's
paper were published under other names, one genus,
Josephia, as named after Banks in the paper, although when
the paper was published it had been changed to *Dryandra*,
after Dryander, who was then just dead.

Goodenough, Bishop of Carlisle, writes to his friend Smith
a little later: " How shocked was I to see Salisbury's sur-
reptitious anticipation of Brown's paper on the New Holland
plants, under the name and disguise of Mr. Hibbert's gardener: Oh, it is too bad! I think Salisbury is got just where
Catline was when Cicero attacked him, viz. to that point of
shameful doing when no good man could be found to defend
him. I would not speak to him at the anniversary of the
Royal Society."

Smith scribbled in his copy of the *Paradisus Londinensis*
the epigram:

> " What malice lurks beneath this fair disguise,
> Satan once more steals into Paradise.
> But now how plausible so'er his tale is,
> We always take his words *cum grano salis*."

Salisbury's carefully preserved herbarium came ultimately to Kew, where he is more strikingly represented by the fine Corsican Pine, near the main entrance, which he procured in 1814. His excellent analytical drawings and notes, which are in the Botanical Department of the British Museum, show him to have been a botanist of more than ordinary ability.

So slavish was then the submission to the Linnæan method, that we read of Dr. George Shaw (1751–1813), Vice-President of the Linnean Society and Keeper of Natural History in the British Museum, being found going through the collection of shells with a small hammer to break all that did not fit the diagnoses of Linnæus. When, too, John Edward Grey (1800–75) stood for election to the Linnean Society, after the publication of his father's *Natural Arrangement of British Plants* (1821), which was understood to be his work, he was blackballed!

CHAPTER XXXVI

ROBERT BROWN

ROBERT BROWN's work is so connected with the advance of Botany in various directions that we have already had several occasions to mention him (Plate XVI).

He was born at Montrose in 1773, his father being an Episcopalian minister. Educated at Aberdeen, and at Edinburgh under Rutherford (the uncle of Sir Walter Scott, who succeeded Hope in 1786), he was appointed, in 1795, surgeon to the Fifeshire Fencibles, then stationed in the north of Ireland. As an extensive collector, he from the first adopted the excellent practice of studying his plants at the time. It is remarkable that, like two other great British botanists of the nineteenth century, W. J. and J. D. Hooker, his first original work was devoted to Mosses. The discovery of *Glyphomitrium Daviesii* introduced him to Banks, who recommended him as naturalist to the Australian coast survey under Flinders, and he sailed in H.M.S. *Investigator* in 1801, having with him Ferdinand Bauer as artist, while one of the midshipmen was John Franklin. The *Investigator* was pronounced unseaworthy, and Flinders started for England for a new ship, but was first wrecked in Torres Straits and then taken prisoner by the French at Mauritius. Brown and Bauer remained in Australia until 1805, however, exploring the Blue Mountains, Tasmania, and the island in Bass Straits, and then returned, after all, in the *Investigator*, bringing back a collection of some 4000 plants, most of which were new to science.

On his return, Brown became librarian to the Linnæan Society, an office that he held until 1822. On Dryander's death, in 1810, he became also librarian to Banks, who—

as we have seen—bequeathed to him his house in Soho Square (where the Linnean Society held their meetings for many years), and the use of his collections, which were to pass to the British Museum on the legatee's death. Brown drove a clever bargain with the Museum Trustees, however, by which the collections were transferred in 1827, he himself becoming an " under-librarian " or Keeper of a newly-constituted Botanical Department, holding office for life at a good salary. He acted as President of the Linnean Society from 1849 to 1853, and died, in 1858, in the library in Soho Square, where he had worked for more than fifty years.

In 1810 appeared the first volume, the only one published, of his *Prodromus Floræ Novæ Hollandiæ*, containing the description of about 200 species in 464 genera, nearly a third of which latter were new to science. Several Families, such as *Santalaceæ*, are here established for the first time; and, as in most of his systematic works, many important anatomical observations are incidentally introduced. His employment of the Jussieuan system of classification was the first signal of revolt in England against the Linnæan tyranny of Smith and his followers.

On Dryander's death in this year, the task of preparing the three remaining volumes of W. T. Aiton's edition of the *Hortus Kewensis* (5 vols., 1810–13), from Dodecandria onwards, devolved upon Brown; and many new genera are described by him in its pages.

Already, in the paper on the Proteaceæ, read before the Linnean Society, to which we have alluded, he had emphasised the study of floral development, and begun the discussion of the essential organs of *Asclepiadaceæ*, which he separated from the *Apocynaceæ*. From this time onwards he was constantly recurring to such questions as the development of the ovule and its coats, the methods of fertilisation and the formation of the embryo, the endosperm and the perisperm.

In 1814 Flinders' *Terra Australis* appeared, with a detailed account of Australian vegetation by Brown, illustrated by

ROBERT BROWN (1773-1858).

facing p. 264.

Plate XV.

Bauer. This contains, in addition to much valuable descriptive botany, the first example of that statistical and geographical method of discussing his results that he afterwards applied to the work of various other travellers. In this he was ably followed by Sir Joseph Hooker, especially in the *Flora Antarctica*, and later by Dr. W. B. Hemsley.

Of Brown's many separate papers none is more important than that of 1827, to which he appended " Observations on the Structure of (the) unimpregnated Ovulum, and on the female flower of Cyadeæ and Coniferæ." In this he summarises all that was known of the ovule, insists on the necessity of studying its development, describes the origin of the seed-coats, distinguishes between perisperm and endosperm, citing the *Nymphæaceæ* as an illustration, and finally demonstrates the gymnospermy of *Cycads, Conifers,* and *Gnetaceæ,* thus paving the way for Wilhelm Hofmeister's work of 1851, which definitely separated the Gymnospermia as one of the main divisions of Phanerogamia.

Although he published no general scheme of classification, Brown was the first to propose the combining of Jussieu's " Orders " into groups of higher value but subordinate to his fifteen " Classes," thus suggesting the Cohorts of Lindley and Bentham and Hooker and the " Orders " of Engler and other recent systematists. In this way he carried a stage further the diagnosis of successive groups of affinity—begun by Bauhin, Tournefort, and Antoine Laurent de Jussieu—and ranks with Fay and Jussieu as one of the chief exponents of the Natural System.

If he was outstripped by Amici in demonstrating the entrance of the pollen-tube into the micropyle, at which Morland long before had guessed, Brown clearly perceived that the germination of the pollen was due to that stimulus of the stigmatic secretion that we now know as chemiotaxis. It was while investigating this matter that he observed the remarkable molecular motion of minute non-living particles since known as " Brownian movement."

It is, perhaps, not generally recognised that it was largely

Brown's studies of floral symmetry and of suppressed parts—
as in Orchids—that not only assisted Darwin in his study of
pollination methods, but led the way to the work of Eichler.
Few botanists nowadays remember that he not only suggested
that *Ophrys* is pollinated without the aid of insects, but also
that the insect-like forms of the labellum or lip of the flower
in that genus are intended, not to attract, but to deter
insect visitors.   It was during his examination of the floral
organs of Orchids that Brown made his chief cytological
observation, the discovery of the nucleus of the cell.   Though
he only partially recognised the significance of this body in
the life of the cell, this was an important step towards the
work of Hugo von Mohl.

The far-reaching import of Brown's work seems to have
been better realised in Germany than at home.   During
one of his vacations he was given the distinguished Prussian
decoration *Pour la Mérite*; and his miscellaneous papers were
published in a collected German edition forty years before
they were so issued in English.   He had, as Sachs says
" found opportunity to leaven the ideas, which through
Humboldt's influence had become predominant respecting
the geography of plants, with the spirit of the natural
system "; and it was Humboldt who bestowed upon him
the title of " *Facile Botanicorum princeps Britanniæ gloria et
ornamentum.*"

CHAPTER XXXVII

PLANT-COLLECTING IN THE TROPICS, ITS SUGGESTIONS AND ITS DANGERS

THE dominating influence of Linnæus upon British and German botanists from about 1760 to 1825 had undoubtedly an impoverishing effect upon botanical science. The explorers of new lands brought back numerous plants, and botany seemed to consist merely in the naming and terse diagnosis of these discoveries. Even when the Natural System struggled to the front, fostered in England by Brown, Gray, and Lindley, the prudent caution of De Candolle in declaring anatomy the only trustworthy guide to affinity did but perpetuate the narrowing of botanical study to the more obvious external characters of plants. As the histology of Grew was left undeveloped and the physiology of Hales aroused the emulation of no disciple, so the most suggestive side of Robert Brown's life-work had but little effect in his own country. There were but few teachers, such as Arthur Henfrey (1819–59), capable of taking a comprehensive view of the entire science and thus appreciating it at its true value. The two Hookers, father and son, were, by the necessities of their official position at Kew, mainly occupied—as was also Bentham—in descriptive work.

The same causes that brought about this stagnation in the science of pure botany, however, resulted in the development of a new and valuable department of what we may term mixed botany—that of plant geography.

We have seen that at the Renaissance it was soon recognised that we must not expect to find in Northern Europe all the species described by Theophrastus or Dioscorides in the south. We have mentioned the solitary passage in which Lobel alludes to the identity of the mountain species of warm regions and those growing at lower levels farther north. Linnæus not only sent his pupils to study the floras of distant lands, but also discusses habitats or " stations " much as do the ecologists of to-day. He insists on soil, aspect, and other topographical conditions, rather than on such broader geographical factors as temperature or altitude. He also admirably discusses the means by which plants disperse themselves. The brilliant imagination of Buffon, and the laborious industry of Adanson, arrived at about the same time at the importance of latitude and altitude. The first adequate conception of the scope of plant geography, however, seems to have been that of the generally ignored Abbé Soulavie (1752–1813), whose two chief works, *La Géographie de la Nature* and *Histoire naturelle de la France meridionale*, were published in 1780 and 1780–4. Soulavie himself states that it was when taken as a young man to Lescrinet, in 1774, for the benefit of mountain air, that he received from his mother and his aunt the first suggestion of altitudinal zones of plant-life. In his later work he describes himself as examining the stations of each species " barometer and thermometer in hand."

Although Carl Ludwig Willdenow (1765–1812) did not become Professor of Natural History in his native city of Berlin until 1798, his *Grundriss der Kräuterkunde* was published there in 1792. Among the aphoristic remarks on geography in this work we read:—" Soil, situation, cold, heat, draught, and great moisture are all of powerful influence upon vegetation." " The warmer the climate, the greater the number of plants." " Those peculiar to polar regions and high mountains are low, have very small compressed leaves, and often in proportion very large flowers." " In cold climates a great number of cryptogams are found, *Cruciferæ*, *Umbelliferæ*, and *Compositæ*, but few trees or shrubs. In warm climates trees, shrubs, ferns, twining plants, parasites, lilies, and *Scitamineæ* abound." "In northern countries, where the cold is so great that no trees can grow, we find beds of coal which are certainly of vegetable origin. In those countries, therefore, forests certainly were once in abundance." "Countries now separated by the ocean were in former times most probably joined. . . . Thus Australia may have been joined to the Cape of Good Hope."

A translation of this work appeared at Edinburgh in 1805, and it would be interesting to ascertain whether in their student days, twenty to thirty years later, it may not have come under the notice of Charles Darwin, Hewett Watson, or Edward Forbes. There can be no doubt that the original was known to Humboldt as soon as it was published.

Friedrich Heinrich Alexander von Humboldt was born at Berlin in 1769. In one of his letters he says: "Of the science of botany I never so much as heard till I formed the acquaintance in 1788 of Herr Willdenow, a youth of my own age."

He stood godfather to one of Willdenow's sons in 1793, dedicated to him his early work, the *Floræ Fribergensis Specimen*, and frequently wrote to him from America. Entering the University of Göttingen in 1789, Humboldt made the acquaintance of Johann Georg Forster, who had accompanied Cook on his second great voyage (1772–5). Forster's account of Cook's voyage is, says the late Dr. Garnett, "almost the first example of the glowing yet faithful description of natural phenomena which has since made a knowledge of them the common property of the educated world—a prelude to Humboldt, as Humboldt to Darwin and Wallace."

This acquaintance confirmed Humboldt's innate love of travel, and he tells us that as early as 1790 he had told Forster of his intention to prepare a work on plant geography. After elaborate training he started for South America in 1799, with his chosen botanical assistant Bonpland, travelling by way of Marseilles, Madrid, and Corunna. They spent six days at Teneriffe, where they met Broussonet and ascended the Peak.

Pierre Auguste Broussonet (1761–1807), a native of Montpellier, who had made the acquaintance of Smith and Banks years before, had acted as deputy professor in the Collège de France. He had become a member of the National Assembly and had been compelled under the Convention, as a Girondist, to quit France. At Teneriffe in 1798—the year before the arrival of Humboldt and Bonpland—he had sketched out a division of the mountain vegetation into four altitudinal zones corresponding to Equatorial, Tropical, Temperate, and Glacial latitudes. This sketch was shown to Humboldt, and must have been a main suggestion for his similar subdivision of Chimborazo.

The main result of Humboldt's three years of exploration in Tropical America was, perhaps, the recognition of isothermal lines and the statistical estimate of the altitudinal and latitudinal distribution of the Natural Orders. It was indeed fortunate that most of Humboldt's travel was tropical travel. " The Tropics," it has been well said, " are biological headquarters." Whilst Europe is, so far as its vegetation is concerned, a " creation of the Glacial Epoch," the Tropics " represent in our cooled and degenerate world the circumstances under which plant and animal life first arose." Thus whilst tropical travel impressed upon Humboldt the general laws of the vertical and horizontal distribution of plant life, it was tropical travel that was to give us at the hands of Darwin and Wallace the theory of Natural Selection. It is an interesting linkage in the history of scientific travel that Humboldt's *Personal Narrative* aroused the enthusiasm of Charles Darwin during his last year at Cambridge, before he seized the opportunity of the voyage in the *Beagle*. Also that it fired Alfred Russel Wallace with the same desire to visit the Tropics. Darwin's own *Journal of a Naturalist* was the companion of the young Joseph Hooker when starting on the *Erebus* and *Terror* expedition in 1839.

Did space permit we might say a good deal as to the dangers of plant-collecting, especially in the Tropics; of the toll of promising young lives that has been paid for the Orchids and other choice plants that enrich our collections, but come in many instances from malarious lands. We must leave this subject, however, with the mention only of David Douglas (1798–1834). The son of a stonemason, he was born at Scone in Perthshire, and, after an apprenticeship in Lord Mansfield's garden, found his way to the Botanical Garden at Glasgow, then under the control of Dr. W. J. Hooker. Thus it was that he was despatched on the first of those collecting expeditions, mainly in North America, on behalf of the Royal Horticultural Society, to which he owes his fame. Among the many new plants that he brought home, *Ribes sanguineum* is now familiar even in the humblest gardens; but the grand Douglas Spruce or Oregon Pine (*Pseudotsuga Douglasii*), represented by the lofty flagstaff at Kew, is more directly associated with his name. Sent out to the Sandwich Islands in 1834, he must have fallen into a pitfall while collecting, for his mangled body was discovered gored to death by a wild bull.

Danger of a different kind, though happily less fatal in its outcome, was that which Joseph Hooker encountered on his second expedition, that to the Himalayas in 1848–50. Personal animosity of a faction at the Court of the Rajah of Sikkim against a fellow-traveller led to the violent detention of Hooker and his companion for two months in 1849. The incident, in fact, nearly ended in one of those assassinations that have so often resulted from trivial misunderstandings in semi-savage countries.

We wish to conclude by emphasising the fact that it was their experience of life in the Tropics, with the relentless struggle of rapidly developing animal and vegetable forms of food, light and air—coupled in the case of both with the re-reading of Malthus's *Essay on Population*—that suggested alike to Charles Darwin and to Alfred Russel Wallace (many years his junior) the great theory of Natural Selection.

One o fthe most romantic stories in the history of science is that of Darwin's receipt from Wallace, after he had already himself devoted years to the elaboration of his theory, of the paper setting forth the same conclusions; of the reading of the summaries of their work on the same evening, July 1, 1858, at the Linnæan Society, by the advice of their mutual friends Lyell and Hooker; and of the chivalrous harmony that marked the subsequent working-out of their results by both naturalists.

CHAPTER XXXVIII

CONCLUSION

In the foregoing chapters we have outlined plant knowledge as it has been gained through the ages by the devoted labours of men responsive to the voice of nature and fascinated with the mystery of life. They gleaned from the open fields of Nature, knowledge of the life that clothed those fields with beauty and made them fit for the pleasure and service of man.

We have brought this story of slowly acquired plant knowledge down to the nineteenth century, and we have outlined the careers of all the leaders in botanical science down to the moderns—such as Paxton and Loudon—who combined the abstract in plant knowledge with the applied in plant culture. With the middle of the nineteenth century the story of the Pioneers of Plant Study closes and the story of the Pioneers of Applied Botany becomes the feature of the world of plants.

The earliest studies of plants, as we have shown, were economic or medical. The cultivation of plants for their own sake, apart from any real need for them, argues culture and refinement worthy of true humanity. The cottager grows his cabbages and potatoes because he needs them; he also grows a few cherished flowers because he loves them. So long as its pioneers were secluded in schools and colleges, Botany played only a minor part in human history. When at last it emerged from its scholarly seclusion and gave humanity a helping hand towards the realisation of the beauties and wonders of nature—then Botany became a world power.

In the development of modern plant knowledge there are

finger posts marking the way. Parkinson's seventeenth-century paradise was premature, yet epoch-making. England, poor and swept by revolution, was not then ready for the modern paradise of the English garden. From the earliest ages all great nations have lavished wealth upon their gardens whenever their riches rose beyond supplying the necessities of life. England had little of such wealth in the seventeenth century—less even than in Elizabeth's day.

With Dutch William III. there came a change. The Dutch were then in the front rank of culture, and the Dutch garden came to England and supplied something more human than the physic garden, although it never was English. During the eighteenth century took place the evolution of the English country garden. This was not a mere enclosure for botanical examples, still less was it a cabbage patch. It embraced the entire landscape and gave the botanist a range impossible in either garden or wild alone.

In the nineteenth century the wealth and power of the Empire furnished the English garden as a garden never before had been furnished in history, and new men and new conditions came into the world of plants. Botany then at last came into its own—not in school, in university, or even in botanic garden, but as handmaid of a human passion and creator of a great industry. It brought to the slum its cherished window flowerpot; to the cottage its flower border just saved from the all-important vegetables; and to the wealthy, gardens of surpassing loveliness.

The educated intelligence of to-day demands the full story of those who, in our century and in the nineteenth, have realised more than Parkinson's paradise. The cottager, educated as his fathers were not, wants the story of his tiny garden. In bleak winter an Indian shrub blooms on his wall and he gives it an Arabic name; a little later, perhaps, it will be outshone by a Japanese beauty. Meanwhile a flower of the Caucasus is struggling over a corner with white bloom, and a Siberian peeps through with deep blue. Then comes another from the uplands of Asia Minor. A primrose too, but its sweet yellow is modest indeed beside the gorgeous Chinese primrose he has tended on his window all winter. With advancing spring, flowers of Persia appear, some with names commemorating great botanists. So on throughout the year there is little that is really English in even the poor cottage garden.

How came these visitors from far-off regions? The reply lends distinction to even the humblest plant as we trace its story through the ages. The beginner in botany chafes at the puzzling nomenclature of plants, yet this conceals a wonderful romance. Some plant names are indeed but modern fancy, but others take us back through the ages to classic times. Bound up with these names are the life stories of the pioneers of plant study, the fruit of whose labours we reap to-day.

INDEX

This exhaustive Index, which has been prepared specially with a view to facilitating reference, includes the names of all persons mentioned in the book.

Where not otherwise stated, the dates are those of the death of the subject of notice, as this is more approximate to his working period than the date of his birth. The plants mentioned are chiefly those of historic interest.

Nestorian, churches and Pepper at Malabar, origin and culture, 84

Nettle, Stinging and Dead, 77

New Learning, the, and plants, 133

New World, the, 114

Nicander, 135 B.C., his 125 plants, 68; *Alexipharmaca* and *Theriaca*, 141; Book of Treacles, 68

Nicobar Islands, 90

Nile, Canopic or Rosetta, mouth of, 50

Nodes, 6

Nomenclator Classicus, Ray's (1675), 209

Nomenclature, scientific and international, 206

Nyctitropism, 63

Oak, evergreen, 74; *Quercus pseudo-coccifera*, 34

Oats, in ancient Egypt, 17

Odoric Friar, his book (1330), 111

Odyssey, plants referred to, 47; fruits of the, 48

Old Testament plants, their eight classes, 32

Oleander, 62

Oleander of Ulysses, 47

Olive, 47; its food value, 34; *Olea europœa*, 32

Onions, 47

Opopanax, an aromatic mentioned by Dioscorides, 91

Orchid (*Roccella tinctoria*), DC., a dye, 114

Oregon Pine, 271

Ornithologia, Willughby, 209

Orta, Garcia (1570), 154

Ourang-outangs of Sindbad, 92

Oviedo, Gonzalo de (1478–1557), Natural History of New World, 119

Oxus, 109

Palæobotany, 1

Palm-wine, 90

Pamir, 109

Papyrus of ancient Egypt, the first paper, 22; brought to Marseilles, 84

Paradise, origin of the word, 128

Paradisi in sole paradisus Terrestris, 186

Parchment, first made at Pergamos, 68

Paris, Botanic Garden (1570), 146

Parkinson, John (1567–1650), apothecary, 181, 188; his memorial, 190; "Park-in-sun's" Earthly Paradise, 188; *Botanicus Regius*

Primarius, 188; first book when sixty-two, 188; book of the garden, 189; *Theatrum Botanicum* and its 3,800 plants, 189

Parkinson, Sydney, 247

Parsley, 96

Pavia, 115

Pea, in ancient Egypt, 17

Pellicier, William, sixteenth century, 168

Pemptades, a Flemish flora, 167

Pena, 167, 169, 239; and Lobel publish *Stirpium Adversaria Nova* in London, 168; visit England (1566), 168

Penny, Dr. Thomas (1589), 168

Pepper, 76, 91, 92; and Nutmeg, Cloves in Borneo, 110; and Cloves, (sixteenth-century cost), 153

Perfumes and spices, 36

Persian Gulf, and pearls, 76

Peru, Tobacco and Coca, 119

Peruvian plants: Banana, Maize, Potato, Tobacco, Agave, Coca, 119

Petiver (1718), apothecary and great collector, 213

Petty, William, 195

Phanerogamia, 3

Pharmacopœia, mediæval, 87; first Government, 141

Philosophia Botanica, 237

Philosophical Transactions (first published in 1665), 198

Phenology, 15

Phœnician, history, 40; enterprise in the Mediterranean, 41; origin of name, 33

Photosynthesis, 7

Phragmites communis, 47

Phytogeography, 1

Pierre Pena, associated with Lobel, 167

Pimpernel, 80

Pineapple (*Ananssa sativa*), 118

Pines and Firs, 3

Pinus Austriaca, 47; *halepensis*, 47; its use in embalming, 22

Pistachio-nut, 35

Pizarro and Peru, 119

Plane and Sycamore, 34

Plane (*Platanus orientalis*), 47

Plantain in Andaman Islands, 110; Museum in Antwerp, 173

Plantarum Umbelliferarum Distributio, Morison, 207

Plant collecting, Fabulous tales of difficulties, 66; plant geography,

288   INDEX

Aristotle and Father of Botany, 60;
his work and books, 60; causes of
plants, 61: history of plants, 61;
450 plants, 62; plant classification,
63; plant descriptions compared
with modern, 64; editions of his
works, 65
Thevet, André (1502–1590), book on
America, 123
Thorns and Thistles, 31
Thyine wood of the Bible, 62
Tibet, 109
Tobacco, 123; *Nicotiana tabacum*,
its history, 117
Tournefort, Joseph Pitton de (1656–
1708), 162, 211, 240; his career,
240; indebted to Cæsalpinus, 187;
Flora of Paris, 241; deprecates
Morison, 208; classification non-
sexual, 240; *Élémens de botanique*
(1694), 240; his 10,000 plants, 74;
and Bernard de Jussieu, 240
Trade routes of the ninth century, 90
Tradescant, John, 182, 190
Trading voyage, first century A.D., 75
Tragus, first botanist to describe with-
out illustrations, 138
Transpiration, 7
Trapezus, now Trebizond, 50
Travels of Belon and his books, 154;
of Marco Polo, 109
Treacle, derived from Greek, 68
Treveris, *Great Herbal*, 129
Trew, Dr., of Nuremberg, 234
Tropical plant collecting, 271
Truffle, 62
Tulip, introduction, 78
Turks, their conquests, 153
Turner (1510–1568), botanist; Gar-
dens at Kew, Wells and Cologne,
145, 147; his travels and *Herbal*,
148; his second botanical work,
148; his persecution, 149; his
Herbal (1551), 149; his *Herbal*
published at Cologne, 150; Dean-
ery of Wells, 149; at Syon House,
245

Vaccinium Oxycoccus, 143
Vaillant (1669–1722), opposed to sex
in plants, 221, 225

Valerand, Dourez, 168
Valerius Cordus, 65; his *Dispen-
satorium*, 141
Varro (28 B.C.) gives 107 plants, 71
Vasco da Gama's voyage to India,
151
Vegetable physiology, 227; *Vegetable
Staticks*, Hales, 228
Vegetative and reproductive, 4
Venice, 126; Marco Polo (1295), 109
Veratrum Album, 53
Vine, 34, 47, 56; kinds known in
A.D. 1, 71
Virgil (70–19 B.C.), his 164 plants, 71
Volga, and ancient trade, 91

Wallace, Alfred Russel, biologist, 271
Walnuts, 35, 93, 100
Water-melon (*Citrullus vulgaris*), 36
Weydiz, illustrator of Brunfels' work,
134
White, Gilbert, *Natural History of
Selborne*, 228
Wilkins, Dr. John, His *Essay on
Character and Philosophy*, 205
Willdenow, Carl Ludwig (1765–1812),
268
William of Orange, and Lobel, 174
Willow, 47
Willughby, Middleton Hall, War-
wickshire, 209; his *Ornithologia*,
209
Withering, William, 258
Woad, 101
Wormwood (*Artemisia*), 32
Wotton, Edward, zoologist, 149
Wren, Christopher, 196
Wyer, Robert, printer, 130

Xerophytic, 8

Yarkand, 109
Yuca (*Manihot*), 181
Yucca Gloriosa, 181
Yunnan, 109

Zouch, Lord, his garden at Hackney,
174